Structural Design for Architecture

For Pat

Structural Design for Architecture

Angus J. Macdonald

Architectural Press

Architectural Press
225 Wildwood Avenue, Woburn, MA 01801-204
An imprint of Butterworth-Heinemann
Linacre House, Jordan Hill, Oxford OX2 8DP
A division of Reed Educational and Professional Publishing Ltd

Ⓡ A member of the Reed Elsevier plc group

OXFORD   BOSTON   JOHANNESBURG
MELBOURNE   NEW DELHI   SINGAPORE

First published 1997
Reprinted 1998

British Library Cataloguing in Publication Data
Macdonald, Angus J.
  Structural design for architecture
  1. Architectural design   2. Structural design
  1.Title
  721

ISBN 0 7506 3090 6

Library of Congress Cataloguing in Publication Data
Macdonald, Angus, 1945-
  Structural design for architecture/Angus J. Macdonald.
    p.   cm.
  Includes bibliographical references and index.
  ISBN 0 7506 3090 6
  1. Buildings.   2. Structural Design.   3. Architectural design.
  1. Title.
  TH846.M33   97-27237
  624. 1' 771-dc21   CIP

Composition by Scribe Design, Gillingham, Kent
Printed and bound in Great Britain

PLANT A TREE
BTCV
British Trust for Conservation Volunteers

FOR EVERY TITLE THAT WE PUBLISH, BUTTERWORTH-HEINEMANN
WILL PAY FOR BTCV TO PLANT AND CARE FOR A TREE.

Contents

Foreword

Angus Macdonald states that this book is primarily for architects. In my view it is also an extremely good reference book on architectural structures for students and practising structural engineers.

He stresses that buildings are designed as a collaborative task between architects and engineers and that the earlier in the design process this happens, the better the result. Current teaching ideas in many universities are, at last, acknowledging the benefits of joint student working and it has certainly been my experience that close working produces the best product.

The early part of the book covers the history, technology and structural philosophy of numerous buildings and building types and has a very comprehensive review of structural systems with excellent examples of seminal buildings and their structures. It also covers the history of structural material development.

The section on structure in relation to architecture: structure ignored, accepted, symbolised and high tech (i.e. celebrated or expressionist) is apt but contentious and could result in some lively discussion between architect and engineer.

The book then divides into sections on the major structural materials – steel, concrete, masonry and timber. Each of these sections follows a similar pattern and includes properties, advantages and disadvantages, common structural forms, etc.

Structural Design for Architecture is a comprehensive and up-to-date work on the relationship of structure to architecture and will form an extremely useful reference work for both students and practitioners of architecture and engineering. I highly recommend it and look forward to having a copy in our office library.

Professor Tony Hunt
Chairman
Anthony Hunt Associates
June 1997

Preface

The architect who considers him or herself to be an artist, dealing through the medium of built form with the philosophical preoccupations of the age in which he or she lives, is surely engaged in a titanic struggle. One aspect of that struggle is the need to determine building forms which are structurally viable. All artists must acquire mastery of the technology of their chosen medium but few face difficulties which are as formidable as those who choose buildings as their means of expression. The sculptor has to contend with similar structural problems but his or her difficulties are trivial by comparison with those of the architect. The difference is one of scale – the size of a building, compared to that of a work of sculpture, means that the technical hurdle which must be surmounted by the architect is of a different order of magnitude to those which are faced by most other artists.

The structure of a building is the armature which preserves its integrity in response to load. It is a bulky object which is difficult to conceal and which must somehow be incorporated into the aesthetic programme. It must therefore be given a form, by the building's designer, which is compatible with other aspects of the building's design. Several fundamental issues connected with the appearance of a building including its overall form, the pattern of its fenestration, the general articulation of solid and void within it and even, possibly, the range and juxtaposition of the textures of its visible surfaces are affected by the nature of its structure. The structure can also influence programmatic aspects of a building's design because the capability of the structure determines the pattern of internal spaces which is possible. Its span potential will deter-mine the maximum sizes of the internal spaces and its type affects the extent to which the sizes and shapes of the spaces can be varied both within an individual storey and between storeys.

The relationship between structure and architecture is therefore a fundamental aspect of the art of building. It sets up conflicts between the technical and aesthetic agendas which the architect must resolve. The manner in which the resolution is carried out is one of the most testing criteria of the success of a work of architecture.

This book is concerned with structural design for architecture. It complements my previous volume, *Structure and Architecture*, and discusses the selection of structure type, the selection of structural material and the determination of structural form. It deals primarily with the development of the *idea* of the structure for a building – that first stage in the structural design process which is concerned with the determination of the elementary form and arrangement of the structure, before any structural design calculations are made. It is intended primarily for architects and it is hoped that it will enable students and members of the profession to gain a better understanding of the relationship between structural design and architectural design. The basic structural layouts and approximate element sizes which are given in Chapters 3 to 6 should, however, also allow building designers to use the book as an aid to the basic planning of structural forms.

Angus Macdonald
Edinburgh
July 1997

Acknowledgements

I would like to thank the many people who have assisted me in the making of this book. These are too numerous for all to be mentioned individually, but special thanks are due to the following: Stephen Gibson for his excellent line drawings, the staff of Architectural Press for their hard work in producing the book, particularly Neil Warnock-Smith, Zoë Youd and Sarah Leatherbarrow. I would also like to thank the staff and students of the Department of Architecture at the University of Edinburgh for the many helpful discussions which I have had with them on the topics covered in this book.

Illustrations, other than those commissioned especially for the book, are individually credited in their captions. Thanks are due to all those who supplied illustrations and especially to the Ove Arup Partnership, George Balcombe, Sir Norman Foster and Partners, Paul H. Gleye, Pat Hunt, Tony Hunt, the late Alastair Hunter, Jill Hunter, Denys Lasdun Peter Softley and Associates, Ewan and Fiona McLachlan, Dr Rowland J. Mainstone and the Maritime Trust. I am also grateful to the British Standards Institution for permission to reproduce tables.

Finally, I should like to thank my wife Patricia Macdonald for her encouragement and support and for her valuable contributions to the preparation of the manuscript and illustrations.

Angus Macdonald

Chapter 1

Structure and architecture

1.1 The role of structure in architecture

The final form which is adopted for a work of architecture is influenced by many factors ranging from the ideological to the severely practical. This book is concerned principally with the building as a physical object and, in particular, with the question of the structural support which must be provided for a building in order that it can maintain its shape and integrity in the physical world. The role of the building as an aesthetic object, often imbued with symbolic meaning, is, however, also central to the argument of the book; one strand of this argument considers that the

Fig. 1.1 Offices, Dufour's Place, London, England, 1984. Erith and Terry, architects. As well as having a space-enclosing function the external walls of this building are the loadbearing elements which carry the weights of the floors and roof. [Photo: E. & F. McLachlan]

Fig. 1.2 Crown Hall, IIT, Chicago, USA, 1952–56. Ludwig Mies van der Rohe, architect. This building has a steel-frame structure. The glass walls are entirely non-structural.

contribution of the structure to the achievement of higher architectural objectives is always crucial. Technical issues are accordingly considered here within a wider agenda which encompasses considerations other than those of practicality.

The relationship between the structural and the non-structural parts of a building may vary widely. In some buildings the space-enclosing elements – the walls, floors and roof – are also structural elements, capable of resisting and conducting load (Fig. 1.1). In others, such as buildings with large areas of glazing on the exterior walls, the structure can be entirely separate from the space-enclosing elements (Fig. 1.2). In all cases the structure forms the basic carcass of the building – the armature to which all non-structural elements are attached.

The visual treatment of structure can be subject to much variation. The structural system of a building can be given great prominence and be made to form an important part of the architectural vocabulary (Fig. 1.3). At the other extreme, its presence can be visually played down with the structural elements contributing little to the appearance of the building (Fig. 1.4). Between these extremes lies an infinite variety of possibilities (see Section 2.2). In all cases, however, the structure, by virtue of the significant volume which it occupies in a building, affects its visual character to some extent and it does so even if it is not directly visible. No matter how the structure is treated visually, however, the need for technical requirements to be satisfied must always be acknowledged. Structural constraints

Fig. 1.3 HongkongBank Headquarters, Hong Kong, 1979–84. Foster Associates, architects. The structure of this building is expressed prominently both on the exterior and in the interior. It contributes directly as well as indirectly to the appearance of the building. [Photo: Ian Lambot. Copyright: Foster & Partners]

Fig. 1.4 Staatsgalerie, Stuttgart, Germany, 1980–83, James Stirling, architect. This building has a reinforced concrete structure and non-structural cladding. Although the structure plays a vital role in the creation of the complex overall form it is not a significant element in the visual vocabulary. [Photo: P. Macdonald]

therefore exert a significant influence, overt or hidden, on the final planning of buildings.

This book is concerned with the programmatic aspects of the relationship between architecture and structure. Chapter 2, in particular, deals with the process by which the form and general arrangement of structures for buildings are determined – with the design of architectural structures, in other words. Information on basic forms of structure – the range of structural possibilities – is essential to the success of this process; this is provided in subsequent chapters which deal separately with the four principal structural materials of steel, reinforced concrete, masonry and timber. More general aspects of the topic are reviewed briefly here.

1.2 Structural requirements

The principal forms of loading to which buildings are subjected are gravitational loads, wind pressure loads and inertial loads caused by seismic activity. Gravitational loads, which are caused by the weight of the building itself and of its contents, act vertically downwards; wind and seismic loads have significant horizontal components but can also act vertically. To perform satisfactorily a structure must be capable of achieving a stable state of static equilibrium in response to all of these loads – to load from any direction, in other words. This is the primary requirement; the form and general arrangement of a structure must be such as to make this possible.

The distinction between the requirements for stability and equilibrium is an important one and the basic principles are illustrated in Fig. 1.5. Equilibrium occurs when the reactions at the foundations of a structure exactly balance and counteract the applied load; if it were not in equilibrium the structure would change its position in response to the load. Stability is concerned with the ability of a structural arrangement which is in equilibrium to accommodate small disturbances without suffering a major change of shape. The first of the beam/column frameworks in Fig. 1.5 is in a

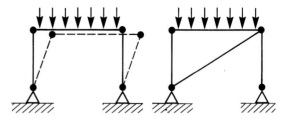

Fig. 1.5 The first of the frameworks here is capable of achieving equilibrium under the loading shown but is unstable. The insertion of the diagonal element in the second framework renders it capable of achieving stable equilibrium.

state of static equilibrium but is not stable and will collapse if subjected to a small lateral displacement. The insertion of a diagonal bracing element in the second framework prevents this and renders the system stable. Most structural arrangements require bracing for stability and the devising of bracing systems is an important aspect of structural design.

As the simple diagrammatic structure in Fig. 1.6 illustrates, the structural elements of a building provide the link between the applied loads and the foundation reactions in order that equilibrium can be achieved. To be effective the elements must be of adequate strength. The strength of an element depends on the strength of the constituent material and the area and shape of its cross-section. The stronger the material and the larger the cross-section the stronger will be the element. It is possible to produce a strong element even though the constituent material is weak by specifying a very large cross-section.

In the case of a particular structure, once the requirements for stability and equilibrium have been met, the provision of elements with adequate strength is a matter firstly of determining the magnitudes of the internal forces which will occur in the elements when the peak load is applied to the structure. Secondly, a structural material of known strength must be selected and thirdly, the sizes and shapes of cross-sections must be chosen such that each element can safely carry the internal force which the load will generate. Calculations are

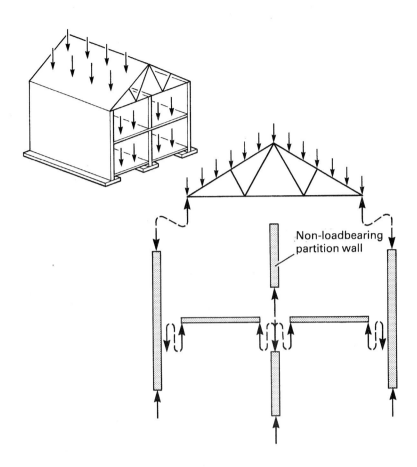

Fig. 1.6 Force system in a building's structure. The gravitational load on the roof is conducted, via the roof truss and the walls, to the foundations where it is balanced by reactions from the substrata. The same is true of loads imposed on the floors which are transmitted by the floor structural elements and walls to the foundations. The roof truss, wall and floor elements must be strong enough to carry the internal forces generated by the load.

Non-loadbearing partition wall

an essential aspect of this process and are required both to determine the magnitudes of the forces in the individual elements – an activity known as structural analysis – and then to calculate the required sizes of the element cross-sections.

A fourth property which a structure must possess, in addition to the requirements of equilibrium, stability, and strength, is adequate rigidity. All structural materials deform in response to load and it is necessary that the overall deflection of a structure should not be excessive. As with strength, the rigidity of the structure depends on the properties of the material and the sizes of the cross-sections, which must be large enough to ensure that excessive deflection does not occur. Like strength, rigidity is checked and controlled through the medium of calculations.

To summarise, the basic requirements of the structure (the firmness element of the architectural shopping list of 'firmness', 'commodity' and 'delight') are the ability to achieve *equilibrium* under all possible load conditions, geometric *stability*, adequate *strength* and adequate *rigidity*. Equilibrium requires that the structural elements be properly configured, stability is ensured by the provision of a bracing system; and adequate strength and rigidity are provided by the specification of structural elements which are of sufficient size, given the strengths of the constituent materials.

1.3 Structure types

1.3.1 Post-and-beam structures
Most architectural structures are of the post-and-beam type and consist of horizontal spanning elements supported on vertical columns or walls. A characteristic of this type

5

Fig. 1.7 Steel skeleton framework. In this arrangement, which is typical of a multi-storey steel-frame structure, concrete floor slabs are supported by a grid of steel beams which is in turn supported by slender steel columns. These elements form the structural carcass of the building. External walls and internal partitions are non-structural and can be arranged to suit planning and aesthetic requirements.

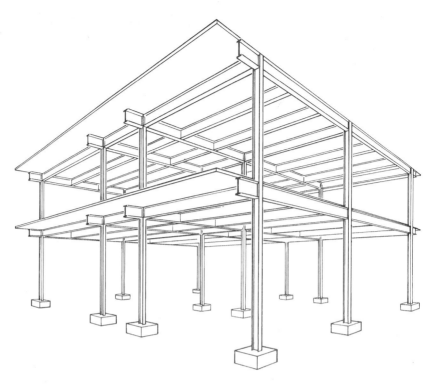

of structure is that the horizontal elements are subjected to bending-type internal forces under the action of gravitational load (normally the primary load on an architectural structure). This has two consequences. Firstly, it requires that the structural material be capable of resisting both tension and compression (e.g. steel, reinforced concrete, timber). Secondly, it is an inefficient type of structure (larger volumes of material are required to support a given load than are necessary with other types of structure).[1] The post-and-beam structure has the great advantage that it is simple and therefore cheap to construct. This group of structures can be subdivided into the two categories of 'skeleton-frame' structures and 'panel' structures. The latter are also loadbearing-wall structures.

Skeleton-frame structures consist of a network of beams and columns which support floor slabs and roof cladding and to which wall

cladding is attached (Fig. 1.7). The configuration of the beam-and-column grid which is adopted in a particular case depends on the overall form of the building concerned, on the internal planning requirements and on the properties of the particular structural material which has been chosen – see Chapters 3 to 6.

In this type of arrangement the structure occupies a relatively small volume and this is in fact one of its principal advantages. Considerable freedom is available to the building designer in the matter of internal planning because both the internal partition walls and the exterior walls are non-loadbearing. Large wall-free spaces can therefore be created in the interiors and different plan-forms adopted at different levels in multi-storey buildings. The choice of external treatment is wide. Relatively fragile materials such as glass can be used and little restriction is placed on the locations of doors and windows (Fig. 1.8).

A consequence of the small structural volume of the skeleton frame is that structural loads are concentrated into slender columns and beams and these elements must therefore

1   See Appendix 1 for an explanation of this.

Fig. 1.8 Nenfeldweg Housing, Graz, Austria, 1984–88. Gunther Domenig, architect. The structure of this building is a reinforced concrete framework. Its adoption has allowed greater freedom to be exercised in the internal planning and external treatment than would have been possible with a loadbearing-wall structure. [Photo: E. & F. McLachlan]

be constructed in strong materials such as steel or reinforced concrete.

Panel structures are arrangements of structural walls and horizontal panels (Fig. 1.9). The walls may be of masonry, concrete or timber – the last of these being composed of closely spaced vertical elements – and the floors and roof of reinforced concrete or timber – again the configuration with timber is one of closely spaced elements, in this case floor joists or trussed rafters (Fig. 6.39).

Many different combinations of elements and materials are possible. In all cases the volume of the structure is large in relation to skeleton-frame equivalents with the result that the structural elements are subjected to lower levels of internal force. Structural materials of low or moderate strength, such as masonry or timber, are therefore particularly suited to this form of construction (Fig. 1.10).

Panel structures impose greater constraints on planning freedom than skeleton-frame equivalents because structural considerations as well as space-planning requirements must be taken into account when the locations of walls are determined. The creation of large interior spaces is problematic as is the variation of plans between different levels in multi-storey arrangements. The advantage of the panel form of structure is that it is simpler to construct than most skeleton frames and considerably less expensive.

Fig. 1.9 Multi-storey loadbearing-wall structure. In this arrangement the floors and roof of the building are carried by the walls. The structural sections give an indication of the action of the walls in response to gravitational and wind loading. The plan arrangement of the walls must be such as to provide good structural performance and the plan-form must be maintained through all levels. These are constraints which must be accepted during the internal planning of this type of building.

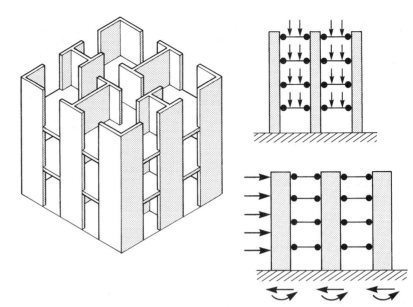

Fig. 1.10 Housing, Gogarloch Syke, Edinburgh, Scotland, 1996. E. & F. McLachlan, architects. These houses are good examples of panel structures in loadbearing masonry. [Photo: Keith Hunter; copyright: E. & F. McLachlan]

1.3.2 Vaults and domes

Vaults and domes are structure types in which the dominant feature is an upwards curvature towards the dominant, downward-acting gravitational load (Fig. 1.11). They belong to a class of structure in which the internal forces are predominantly axial rather than of the bending type,[2] and, in the case of vaults and domes, this internal force is compressive. They are therefore normally constructed in materials which perform well in compression, such as masonry or concrete.

The axial-compressive-stress-only condition which is associated with vaults and domes has two important consequences. First, it allows

Fig. 1.11 TGV Station, Lyon-Satolas, France 1989–94, Santiago Calatrava, architect/engineer. A vaulted structure in reinforced concrete is used here to achieve a relatively long span. [Photo: E. & F. McLachlan]

large horizontal spans to be achieved with materials, such as masonry or unreinforced concrete, which have little tensile strength (Fig. 5.4): large-span interiors can be created in masonry only by the use of domed or vaulted structures. This was the principal reason for the use of this type of arrangement prior to the invention of modern materials such as steel and reinforced concrete which allow large spans to be achieved with post-and-beam forms due to their ability to resist bending effectively.

Secondly, and perhaps more importantly for the buildings of today, it allows loads to be resisted with much greater structural efficiency than is possible where bending is the principal

---

2 See Macdonald, *Structure and Architecture*, Chapter 4 for a classification of structure types. The principles are summarised here in Appendix 1.

Fig. 1.12 Building for IBM Europe travelling exhibition. Renzo Piano, architect/engineer, Ove Arup and Partners, structural engineers. A complex, highly efficient vaulted structure like this would not normally be used for a short-span enclosure. It was justified in this case due to a requirement for a lightweight structure for a portable building. The choice of lightweight materials – timber and plastic – was also sensible. [Photo: Ove Arup & Partners]

result of the application of load.[3] In modern practice, vaults or domes are normally used to achieve high levels of structural efficiency, either because a very long span is required or because a special requirement must be satisfied such as the need for a very lightweight structure (Fig. 1.12).

1.3.3 Tents and cable networks

Tents and cable networks are tensile equivalents of domes and vaults (Fig. 1.13). The internal forces which occur in these structures are those of axial tension and they are there-fore, like their axial compressive equivalents, potentially highly efficient in resisting load.[4] As with domes and vaults they are used in situations in which high structural efficiency is desirable, such as for long spans or where a lightweight structure is required.

1.3.4 Combined-action structures

A fourth category of structure is one in which the load is resisted by a combination of bending and axial internal forces. The ubiquitous portal frame (Fig. 1.14) is perhaps the best-known example of this but any structure which is neither purely 'form-active' nor purely 'non-form-active' will carry load through the combined effect of axial and bending action. These structures have properties which are intermediate between those of the post-and-beam arrangement, which is inefficient but simple to construct, and the arch, vault or cable network, which are highly efficient but

3 See Macdonald, *Structure and Architecture*, Chapter 4.   **4** See Appendix 1.

complicated to construct. Combined-action structures are therefore used in situations in which intermediate levels of efficiency are required, for example in the medium-span range. They are most often found in the form of skeleton-frame arrangements.

Fig. 1.13 Olympic Stadium, Munich, Germany, 1968–72. Behnisch & Partner, architects, with Frei Otto. The structure of this canopy consists of a network of steel wires (the very fine square mesh) supported on a system of masts and cables. The pattern of heavy rectangular lines results from the flexible joints between the cladding panels. Highly efficient structure types such as this are required where long spans are involved. [Photo: A. Macdonald]

1.4 Structural materials

The form and general arrangement of architectural structures are greatly influenced by the properties of the materials from which they are constructed. For this reason the basic structure types appropriate to the four principal materials of steel, reinforced concrete, masonry and timber are described in separate chapters.

Each material has its own individual characteristics in terms of physical properties and manufacture which contribute to determining the structural forms for which it is most suitable. These issues are considered in detail in the chapters on individual materials. Only the most general aspects are reviewed here.

The properties of materials which affect the load-carrying performance of a structure are strength, elasticity and, to a lesser extent, specific gravity (which determines the self-weight of structural elements). Other significant physical properties are durability (i.e.

11

Fig. 1.14 Palmerston Special School, Liverpool, England, 1973–76 (demolished 1989). Foster Associates, architects; Anthony Hunt Associates, structural engineers. Semi-form-active portal frames of steel hollow-section are used here as the primary structural elements in a multi-bay arrangement with relatively short spans. The moderately high efficiency of this type of structure has permitted very slender elements to be adopted. [Photo: John Donat]

susceptibility to both physical and chemical deterioration) and performance in fire. Non-physical, but interrelated, properties which are relevant are cost, availability and environmental impact. The last of these is concerned with the environmental issues (depletion of material resources and energy sources, pollution, the health of workers, etc.) which will arise from the manufacture, installation and use of structural elements of a particular material.

Of the purely physical properties, perhaps the most important so far as structural performance is concerned is strength, although the ratios of strength to weight and strength to elasticity are also significant because these determine the efficiency with which a material can be used. Of the four principal structural materials, steel and reinforced concrete may be thought of as high-strength materials and timber and masonry as low-strength materials. Each of the four has a unique combination of properties which makes it perform best in particular types of structural arrangement.

Another set of factors which influences the types of structure for which a material is suitable are the conditions of its manufacture and finishing. These determine the type of product in which the material becomes available to the builder. Steel, for example, is available in the form of manufactured elements which are straight and of constant cross-section. The construction of a steel structure is therefore a process of assembly of prefabricated components. Concrete, on the other hand, normally arrives on a building site in liquid form and the building is literally formed

Fig. 1.15 The high strength of steel allows the creation of structures with very slender elements. In buildings which are supported by steel frameworks the volume occupied by the structure is low in relation to the total volume of the building. (Photo: A. Macdonald).

Fig. 1.16 Multi-storey steel frameworks are typically a combination of I-section beams and H-section columns. (Photo: A. Macdonald).

on the site by pouring the concrete into moulds.

The strongest of the structural materials is steel, which is therefore used for the tallest buildings and the longest spans. It is a highly versatile material, however, and is also used over a very wide variety of building types and span sizes. Because it is very strong the structural elements are slender and of low volume, so that steel is used almost exclusively in the form of skeleton-frame structures (Fig. 1.15). The majority of these are assembled from standard rolled sections; these are elements with I- and H-shaped cross-sections (Fig. 1.16) and longitudinal profiles which are straight

13

and parallel-sided, and which lend themselves to use in straight-sided frameworks. Most steel structures therefore have a regular, rectilinear geometry. The range of possible forms has been extended in recent years by the development of techniques for bending rolled sections into curved shapes and by the increased use of casting to produce structural steel elements. However, the fact that all steel structures are prefabricated tends to require that regular and repetitive structural geometries be adopted even though the individual components are of irregular or curvilinear shape.

A typical steel-frame building thus has a relatively simple overall form and an interior which is open and unencumbered by structural walls. Great freedom is therefore available to the designer so far as the internal planning of such buildings is concerned: the interior volumes may be left large or they may be subdivided by non-structural partition walls; different arrangements of rooms may be adopted at different levels and a free choice is available in the treatment both of the external walls and of the internal partitions.

An advantage of prefabrication is that steel structural elements are manufactured and pre-assembled in conditions of very high quality control. Great precision is possible and this, together with the slenderness which results from high strength, means that structures of great elegance can be produced. Steel is therefore frequently selected as much for its aesthetic qualities and for the stylistic treatment which it makes possible as for its structural performance.

Reinforced concrete, the other 'strong' material, is of lower strength than steel with the result that equivalent structural elements are more bulky. It too is used principally in skeleton-frame structures of regular geometry and therefore offers similar advantages to steel in respect of internal planning and exterior treatment.

Concrete structures are normally manufactured on the building site by the pouring of liquid concrete into temporary formwork structures of timber or steel which are specially made to receive it. This allows a wide choice of

Fig. 1.17 Goetheanum, Eurhythmeum, 'Glashaus' studio, Dornach, Switzerland, 1924–28. Rudolf Steiner, architect. The complexity of form which is possible with *in situ* reinforced concrete is well illustrated here. [Photo: E. & F. McLachlan]

element shape to be available. Continuity between elements is also easily achieved and the resulting statical indeterminacy[5] facilitates the production of structures of complicated form. Irregular geometries in both plan and cross-section, with cantilevering floor slabs, tapering elements and curvilinear forms may all be produced more easily in reinforced

---

5 See Macdonald, *Structure and Architecture*, Appendix 3, for an explanation of the phenomenon of statical indeterminacy and its relevance to the determination of structural form.

Fig. 1.18 Casa Pfaffli, Lugano, Switzerland, 1980–81. Mario Botta, architect. The structure of this building consists of loadbearing masonry walls supporting reinforced concrete horizontal structural elements. [Photo: E. & F. McLachlan]

concrete than in steel (Figs 1.17 and 4.20). The shapes of the elements are usually, however, more crude at a detailed level.

Masonry is the term for a range of materials which have the common characteristic that they consist of solid elements (bricks, stones, concrete blocks, clay tiles) which are bedded in mortar to form piers and walls. A range of other materials with similar physical properties to masonry, such as various forms of dried or baked earth, are suitable for the same types of structural configuration.

The principal physical properties of these materials are moderate compressive strength, relatively good physical and chemical durability and good performance in fire. Very significant properties are brittleness and low tensile strength. The last of these, in particular, has a profound effect on the structural forms for which masonry is suitable. Lack of tensile strength means that bending-type load of significant magnitude cannot be resisted so that masonry structural elements must be subjected principally to axial compression

only. They can be used as walls, piers, arches, vaults and domes but not as slab-type horizontally spanning elements. When used as walls, they must be supported laterally at regular intervals due to their inability to withstand out-of-plane bending loads such as might occur due to the effect of wind pressure.

Masonry is therefore used in the loadbearing-wall form of structure (Fig. 1.18) to produce multi-cellular buildings in which the principal walls are continuous through all levels, giving similar arrangements of spaces on every storey. The horizontal elements in such buildings are normally of timber or reinforced concrete but may be of steel. Structural continuity between these elements and the supporting masonry walls is difficult to achieve and the internal forces in the structural elements must be maintained at modest levels. Spans are therefore normally kept small and in modern practice loadbearing masonry buildings are usually fairly small in scale. (This is in contrast to the very large-scale masonry structures of previous ages which were achieved by the use of masonry vaults and domes as the horizontal spanning elements – see Section 5.2. and Fig. 5.1.)

Although modest in scale, modern loadbearing masonry structures exhibit very good combinations of properties and produce buildings which are durable and fireproof and which have walls which perform extremely well in respect of thermal and acoustic insulation. They are therefore ideal for all kinds of living accommodation.

Timber is a structural material which has similar properties to steel and reinforced concrete in the sense that it can carry both tension and compression with almost equal facility. It is therefore capable of resisting bending-type load and may be used for all types of structural element. It is significantly less strong than either steel or reinforced concrete, however, with the result that larger cross-sections are required to carry equivalent amounts of load. In practice, large cross-sections are rarely practicable and timber elements must therefore normally be used in situations where the internal forces in the structural elements are low, that is in buildings of small size, and, in particular, short spans.

A significant advantage which timber has over other structural materials is that it is very light, due to its fibrous internal structure and the low atomic weights of its constituent chemical elements. This results in a high ratio of strength to weight. Other advantageous properties are good durability and, despite being combustible, relatively good performance in fire.

Timber is commonly used for the horizontal floor and roof elements in loadbearing masonry structures (Fig. 6.39). Loadbearing-wall 'panel' construction, in which the structure of a building is made entirely of timber, is another common configuration (Fig. 1.19). Wall panels consist of closely spaced timber posts

Fig. 1.19 Timber loadbearing-wall structure. Everything here is structural. The wall and floor structures consist of closely spaced timber elements. Temporary bracing elements, which provide stability until non-loadbearing cross-walls are inserted, are also visible. [Photo: A. Macdonald]

tied together by horizontal timber elements at the base and top, and the panels are arranged in plan configurations which are similar to those used in masonry construction. The grouping together of timber posts into panels ensures that the load which each carries is relatively small. Even so, such buildings rarely consist of more than two storeys. Timber loadbearing-wall structures are simple to construct, using components which are light in weight and easily worked. The use of timber in skeleton-frame configurations for multi-storey structures is relatively rare and normally requires that columns be closely spaced to

Fig. 1.20 Forth Railway Bridge, Scotland, 1882–90, Henry Fowler and Benjamin Baker, engineers. Two types of structure are visible here. Short-span viaducts, consisting of parallel-chord triangulated girders, are used at each end where the ground conditions allow closely-spaced foundations to be provided. The two deep river channels separated by the island are spanned by an arrangement of three pairs of balanced cantilevers. The configuration of the bridge was determined by a combination of site conditions, structural requirements and function. [Photo: P. & A. Macdonald]

keep beam spans short. Very large single-storey structures have, however, been constructed (see Section 6.2).

1.5 Structural design

As with any other type of design, the evolution of the form of a structure is a creative act which involves the making of a whole network of interrelated decisions. It may be thought of as consisting of two broad categories of activity: first, the invention of the overall form and general arrangement of the structure and, secondly, the detailed specification of the precise geometry and dimensions of all of the individual components of the structure and of the junctions between them.

In the case of an architectural structure both of these activities, but especially the first, are closely related to the broader set of decisions connected with the design of the building. The

17

overall form of the structure must obviously be compatible with, if not identical to, that of the building which it supports. The preliminary stage of the structural design is therefore virtually inseparable from that of the building, taken as a whole. It is at this stage that architectural and structural design are most closely related and that the architect and engineer, be they different persons or different facets of the same individual, must work most closely together. The second stage of the design of the structure, which is principally concerned with the sizing of the elements and the finalising of details, such as the configuration of the joints, is principally the concern of the structural engineer.

The different aspects of the structural design activity are most easily seen in relation to purely engineering types of structure, such as bridges. It will be instructive here, before looking at the process as it takes place in the case of a building, to consider the design of a prominent example of this type. The Forth Railway Bridge in Scotland (Fig. 1.20), despite being now over 100 years old, provides a good illustration of the various stages in the evolution of a structural design. The issues involved are broadly similar to those which occur in any engineering design project, including those of the present day. The same issues will be considered again in Chapter 2, where they are discussed in relation to architectural design. This preliminary review, in the context of engineering, serves to identify the essential aspects of the structural design process.

As is normal in bridge design, the most significant sets of factors which influenced the design of the Forth Bridge were those connected with its function and with its location. The ground level at each side of the estuary of the River Forth, where the bridge is situated, slopes steeply up from the shore and the railway therefore approaches from a level of approximately 50 m above water level at each end. This, together with the requirement that the busy shipping channels which the bridge crosses should not be blocked, dictated that the railtrack level should also be relatively high (50 m above sea level). At one shore a broad strip of low lying ground occurs close to the

Fig. 1.21 Forth Railway Bridge, Scotland, 1882–90, Henry Fowler and Benjamin Baker, engineers. The main part of the structure consists of three pairs of balanced cantilevers. In this shot the central tower of one pair of cantilevers has been completed and the first elements of the cantilevers themselves have been added. The arrangement was adopted so that the uncompleted structure could be self-supporting throughout the entire period of construction. [Photo: E. Carey; copyright: British Rail Board Record Office]

edge of the water. At the other, the ground rises steeply from the water's edge but a broad strip of shallow water occurs close to the shore. Between these two flanking strips of relatively level ground the estuary consists of two very deep channels separated by a rocky island. The bridge was therefore broken down into three parts and made to consist of two long approach viaducts, each with a sequence of girders spanning relatively short distances between regularly spaced piers, and a massive central structure spanning the two deep channels.

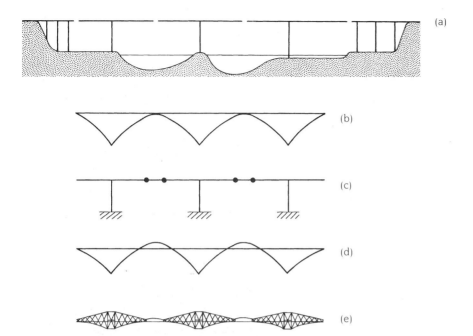

(a)

(b)

(c)

(d)

(e)

The structure which spans the central channels consists of three sets of balanced cantilevers – a configuration which was adopted because it could be built without the need for temporary supporting structures. The arrangement was essential due to the near impossibility of providing supporting structures in the deep channels and due to the need to maintain the shipping lanes free of obstruction during the construction process.[6] The method of construction which was adopted was to build three towers first and then extend pairs of cantilevers simultaneously on each side of these (Fig. 1.21). The structure was therefore self-supporting during the entire process of construction.

The basic form of the structure is shown diagrammatically in Fig. 1.22 together with the bending moment diagram[7] which results from the peak load condition (a distributed load

Fig. 1.22 Forth Railway Bridge, Scotland, 1882–90, Henry Fowler and Benjamin Baker, engineers.
(a) A diagrammatic representation (not to scale) of the main elements of the bridge. The two river channels are crossed by three sets of balanced cantilevers. Viaducts consisting of short-span girders supported by closely spaced piers provide the necessary link with the high ground on the approaches to the bridge.
(b) The bending moment diagram which results from the action of a uniformly distributed load across the entire structure gives an indication of the variation which occurs in the magnitudes of the internal forces across the span.
(c) The insertion of two extra hinges creates a cantilever-and-suspended-span arrangement which alters the bending moment diagram (d) and reduces the magnitude of the maximum bending moment.
(e) The external profile which was finally adopted is closely related to the bending moment diagram. The triangulation of the internal geometry was necessary to achieve high structural efficiency.

6   The device of constructing the main elements of the bridge on the shore and floating these into the final position, which had been used earlier in the nineteenth century at Saltash (I. K. Brunel) and the Menai Straights (Robert Stephenson and William Fairbairn) was impractical due to the very long spans involved.

7   The bending moment is the internal force produced by the load on the structure in the type of arrangement shown in Fig. 1.22. The bending moment diagram is a graph which shows how the intensity of this internal force varies across the span. For an explanation of bending moment see Macdonald, *Structure and Architecture*, Chapter 2.

Fig. 1.23 Forth Railway Bridge, Scotland, 1882–90, Henry Fowler and Benjamin Baker, engineers. The railway track is carried on an internal viaduct which is supported at the junctions of the triangulated main structure. [Photo: A. Macdonald]

across the entire span). This shows that the intensity of internal force is at its highest at the locations of the support towers and falls to zero at the mid-span points, where the adjacent cantilevers are joined. The distribution of internal force was modified by the insertion of two hinge-type connections between each set of cantilevers (Fig. 1.22c) which had the effect of reducing the magnitude of the maximum internal force at each support tower.

Figure 1.22c represents, in diagrammatic form, the basic configuration which was finally adopted for the bridge. It was modified to give improved load-carrying efficiency by matching the longitudinal profile of the structure to the pattern of internal forces so that the structural material was concentrated at the locations of highest internal force. It was further improved by the adoption of a triangulated internal geometry. Yet another decision taken by the designers was to carry the railtrack on a viaduct (similar to the approach viaducts) located within the primary structure and supported by it at regular intervals (Fig. 1.23). The primary structure therefore provided regularly spaced supports for this internal viaduct rather in the manner of the equally spaced piers which carry the approach viaducts.

Figure 1.22e shows diagrammatically the final form of the bridge. It represents the culmination of the first stage of the design process in which the form and general arrangement of the entire structure were determined. The sequence of decisions which led to this proposal illustrates, in a much abbreviated form,[8] the key stages of the preliminary design process and allows the nature of the process which concerns us here to be appreciated. In particular, it shows that the basic form of this purely engineering structure was determined by the designers from a consideration of the function of the bridge, from the constraints of the site, from a knowledge of structural behaviour and from an awareness of the vocabulary of structural possibilities which was available to them. The design was an imaginative response to these various influences.

The second stage in the design process, the realisation of the structure, led to the detailed specification of the various structural elements from which the bridge was constructed. This involved decisions on the precise shape of the individual elements (in longitudinal profile and cross-section) and on the precise amount of material which would be specified (the overall size of each cross-section, thickness of metal, etc.). At this stage structural calculations were employed to determine the magnitudes of the internal forces which each element would carry and to check that sufficient thicknesses of metal were specified to maintain the stresses within acceptable limits. The detailed determination of the configuration of the joints between the elements, which could only be done once the sizes of the elements were known, was then carried out.

The description given above illustrates, in much abbreviated form, the typical sequence of decisions which is involved in the design of any major civil engineering structure and allows the two key stages in this – the initial determination of form and the final realisation of the form – to be appreciated. A sequence of decision-making which is broadly similar to this is involved in the design of all structures whether civil engineering or architectural.

In the case of an *architectural* structure, the form and general arrangement must also be entirely compatible with that of the building which it will support. The preliminary stage in the design of an architectural structure is therefore inseparable from the design process of the building as a whole. Once the overall form of a building has been determined, the choice of structural arrangement will normally be fairly narrow. The act of architectural design is therefore also an act of structural design in which the most fundamental decisions relating to the structure are taken.

---

8 Several alternative arrangements were in fact considered by the designers before the chosen form was adopted.

Chapter 2

Structural design for architecture

2.1 Introduction

The purely technical aspects of the design of a structure were reviewed at the end of the previous chapter. The complex relationship between structural design and architectural design will now be considered.

As was seen in Section 1.5 the process of structural design may be subdivided into two parts: there is a preliminary design stage, when the form and general arrangement of the structure are devised, and a second stage in which structural calculations are performed and the dimensions of the various structural elements are determined. In the case of an architectural structure many of the decisions associated with the preliminary stage of the design of a structure are taken, consciously or unconsciously, when the form of the building is determined. The general arrangement chosen for a building will normally determine the type of structure which will have to be adopted to support it and will probably also dictate the selection of structural material.

In the case of the Willis, Faber and Dumas building (Fig. 4.17), for example, where there was a requirement for a large wall-free interior and glass external walls, there was no alternative to the adoption of a frame-type structure. The requirements for a curvilinear plan-form and for columns which were set back from the perimeter, dictated that reinforced concrete rather than steel be employed as the structural material. The outcome in this case was a building in which architectural and structural requirements were satisfied in equal measure and the building stands up well to both architectural and technical criticism.

Fig. 2.1 Rooftop Remodelling in Vienna, Austria, 1988. Coop Himmelblau, architects. This glazed, irregular form required that a skeleton framework structure be adopted. [Photo: Gerald Zugmann]

The complex arrangements of the Rooftop Office in Vienna by the Coop Himmelblau group, to take another example (Fig. 2.1), would have been unrealisable with any other type of structure than a skeleton framework, which had to be of steel to ensure that the elements were sufficiently slender. The Hysolar building of Behnisch (Fig. 2.2) is a similar type of building and could only have been realised with a skeleton framework of structural steel. The buildings of Richard Meier (Fig. 4.18) and Frank Gehry (Fig. 4.20), on the other hand, required that reinforced concrete structures be adopted.

Thus, although some aspects of the design of structures, such as the precise geometry of a

Fig. 2.2 Solar Research Institute (Hysolar), Stuttgart, Germany, 1988–89. Gunther Behnisch & Partner, architects. The irregular geometry and large areas of glazing required that a skeleton framework structure be adopted to support this building. The choice of steel produced a particular aesthetic quality – the most notable aspects of this are the slenderness of the structural elements and the refined appearance of the exposed steelwork. [Photo: E. & F. McLachlan]

beam or column grid or the dimensions of the elements, may remain undecided until a later stage in the design, many important structural choices will be closed once the overall form of a building has been determined. The initial concept for a building, which determines its overall form and the disposition of solid and void within it, therefore exerts a dominating influence on its subsequent structural make-up. The architect is thus, consciously or unconsciously, a structural designer.

The design of the structure which will support a building is an identifiable and discrete part of the overall design process in

23

which four broad categories of decision-making may be identified. These are: first, the decision on the kind of relationship which will exist between the architectural design and the structural design; secondly, the selection of the generic type of structure for the building; thirdly, the selection of the structural material; and lastly, the determination of the detailed form and layout of the structure. In a particular case these sets of decisions may not be taken in an ordered sequence and some – for example on the relationship which is to be adopted between structural and architectural design – may not even be taken consciously. They are nevertheless decisions which are taken by the designer of every building.

In the design of the majority of buildings, in which the relationship between architectural design and structural design is not one of 'structure ignored' (see Section 2.2 for an explanation of this term), the preliminary structural decisions are taken consciously, usually by a team of designers which includes architects and structural engineers. The nature of the decision-making sequence is explored in this chapter.

2.2 The relationship between structural design and architectural design

2.2.1 Introduction

The detailed design of a structure is normally carried out by a structural engineer (more probably a team of engineers) but, as was noted above, the overall form of an architectural structure is determined by that of the building which it supports and therefore principally by the architect (or architectural team). This raises the issue of the extent to which the architect should be preoccupied by structural considerations when determining the form and general arrangement of a building.

There are several possible approaches to this. Leon Baptista Alberti, who in his treatise on architecture written in the fifteenth century more-or-less outlined the job description of the modern architectural profession, stated:

It is quite possible to project whole forms in the mind without any recourse to material, ...[1]

More recently, Wolf Prix of Coop Himmelblau expressed the view that:

... we want to keep the design moment free from all material constraints ...[2]

It is possible to say therefore that at least some European architects have, since the Renaissance, considered it feasible to ignore structural considerations when they invent the form of a building, believing that a preoccupation with technical matters inhibits the process of creative design. The second of the quotations above is a fairly explicit comment from a contemporary architect whose work many would regard as controversial but it nevertheless describes the reality of a very significant proportion of present-day architectural design activity, including much of what might be thought of as belonging to the mainstream rather than the fringe. Many architects, in fact, pay little attention to structural issues when determining the form of a building. Few, however, find it appropriate to make statements such as that quoted above.

Just as it is possible virtually to ignore structural issues when carrying out the preliminary design of a building, it is also possible to allocate the highest priority to the satisfaction of structural requirements. In such cases the aesthetic and programmatic aspects of the design are accorded a lower priority than those connected with the structure and are not allowed to compromise the quality of the structural design. This approach can occur through necessity, when the limits of what is feasible structurally are approached due, for example, to the extreme height of a building or

---

1 L. B. Alberti, *On the Art of Building in Ten Books*, 1486, Trans. Rykwert, Leach and Tavernor, London, 1988.
2 Quotation from 'On the Edge', the contribution of Wolf Prix of Coop Himmelblau to *Architecture in Transition: Between Deconstruction and New Modernism*, Noever (ed.), 1991.

to the need for a very long span (e.g. Fig. 3.10). It can also occur through choice.

Some architects hold the view that one of the marks of a well-designed building is that all potential conflicts between the architectural programme and its structural consequences have been resolved without either aspect dominating the other. This represents yet another relationship between structure and architecture and it is one with which most mainstream architects would openly, if perhaps somewhat hypocritically, concur. In this scenario well-designed structure is regarded as a necessary precondition for good architecture. Such an approach requires that the structural make-up of a building be evolved in conjunction with all other aspects of its design and that structural issues be considered from an early stage in the design process and allowed to play as significant a part in the determination of the final form of a building as the aesthetic and space-planning programmes. In this approach the objective of design is to determine the form in which all requirements are satisfied equally.

It is possible therefore for structure and architecture to be related in several different ways and one of the first decisions which has to be taken by a design team which has a full awareness of the activity in which it is engaged, is concerned with the nature of this relationship. Often the issue is allowed to remain unclear and some architects even deny that there can be more than one proper relationship between structure and architecture. It has been suggested however[3] that the relationship can take a number of different forms and that the totality of possible relationships between structure and architecture may be summarised in the four categories of *structure ignored*, *structure accepted*, *structure symbolised* and *structural 'high tech'*. The implications of these categories of relationship are now reviewed.

2.2.2 Structure ignored

It is possible, principally due to the existence of structural materials such as steel and reinforced concrete, to invent architectural form without considering the structural implications of that form. The wide variety of forms into which steel is fashioned in the manufacture of motor cars, ships, marine oil production platforms and consumer durables, for example (Fig. 2.3), all of which have shapes which are determined principally from criteria other than those connected with structure, illustrate the almost limitless possibilities which are present in the matter of form when

Fig. 2.3 Steel can be fabricated into almost any shape as the structures which are built for the marine environment demonstrate. The construction of a ship requires techniques of fabrication which are significantly more complex and expensive than those which are normally used in the construction of buildings. There is, nevertheless, no technical reason why such shapes should not be employed in architecture. [Photo: P. Macdonald]

---

3 See Macdonald, *Structure and Architecture*, Chapter 7.

Fig. 2.4 Sailing Yacht 'Gipsy Moth IV'. In the construction of ships and yachts timber has been used to produce a wide variety of complex shapes. The material itself places little restriction on form but the production of geometries like that illustrated is highly labour intensive and therefore expensive. [Photo: The Maritime Trust]

the objects concerned are made from steel. A similar statement could be made of reinforced concrete but in this case the exemplars would perhaps be grain silos or water towers. Timber too is capable of being used in a very wide range of shapes although, because timber is less strong than steel or reinforced concrete, these are of a smaller scale. Again, ships and yachts serve as examples of the level of complexity of form which is possible (Fig. 2.4). In all of these examples a criterion other than structure, such as hydrodynamic performance or adherence to fashion, was the dominating influence on the form which was chosen.

The structural factor which is common to the three materials considered above (steel, reinforced concrete and timber) and which allows them to be shaped into almost any form, is that they can all resist tension, compression and bending.[4] The high strength of steel and reinforced concrete, together with the fact that very effective structural joints can be made between components of these materials, are additional factors which make possible the creation of practically any form.

It may seem surprising therefore that, in the period since these materials became widely available to the builder in the late nineteenth century, architects have availed themselves so little of the potential for the free invention of form which they made possible. Instead, architects have, for the most part, generally continued to produce buildings with plane vertical walls and horizontal or pitched roofs which are not significantly different in overall shape from the traditional architectural forms which were evolved from the much more restrictive structural technology of masonry. There have been several reasons for this, most of which were not technical.

Perhaps the most significant reason for the apparent lack of basic (as opposed to superficial) complexity in the architectural forms of the twentieth century has been cultural. In the period in which the freedom to experiment with form has been available it has not generally been fashionable for buildings to be given irregular or curvilinear forms. For ideological reasons, modernity preferred to accept the vocabulary of orthogonality, which symbolised

---

4 They can therefore be used to make any type of structural element: form-active, semi-form-active or non-form-active. The same cannot be said of masonry, which has very limited tensile strength and therefore also limited bending strength. As is shown in Chapter 5, the need to prevent significant tensile stress from developing imposes constraints on the structural forms of masonry.

Fig. 2.5 East Pavilion, Groninger Museum, Groningen, Netherlands, 1990–94, Coop Himmelblau, architects. Parts of this building were manufactured in a shipyard using shipbuilding techniques. It is a relatively rare example in architecture of the employment of a large-scale industrial process to create a complex form. [Photo: R. Talbot]

rationality; of regularity and repetition, which symbolised the production line; and of straight lines and sharp edges, which symbolised manufacture by machine rather than by hand crafting (e.g. Figs 1.2 and 4.6). Thus, at the very time when developments in structural engineering could have released architects from the tyranny of the straight line and the right angle they voluntarily restricted themselves to the use of little else. In recent years the mood has changed, however, and architects of the late modern Deconstruction school, in particular, are now availing themselves of the freedom which their predecessors failed to exploit.

This may be seen in buildings such as those illustrated in Figs 2.1, 2.2 and 4.20. It is interesting to note that in the East Pavilion of the Groninger Museum (Fig. 2.5), major elements of the steelwork were manufactured in a local shipyard using shipbuilding techniques. This allowed the use of shapes and configurations which, in the context of building, were exciting and new. It is a matter of conjecture whether this can be taken seriously as a method by which buildings should be constructed but it does show that architects are finally making use of the full potential of the materials which industry has placed at their disposal.

One of the reasons for the continuation, into the twentieth century, of the relatively simple structural geometries of architectural tradition was the elementary one of convenience. Arrangements of rectangular rooms with horizontal floors and ceilings are more suitable, for most human purposes, than enclosures made from curvilinear surfaces or wall and roof planes which intersect at acute or oblique angles. There was no technical reason, however, why buildings with complex forms, with non-vertical walls, inclined roof planes and curvilinear enclosures could not have been constructed once materials such as steel and reinforced concrete became available to the designers of buildings in the late nineteenth century.

Yet another factor, which is not strictly technical but which may have inhibited the use of irregular forms, is obviously that of cost. Complicated forms are difficult and therefore expensive to construct. This is why artefacts such as motor cars or aeroplanes are relatively more expensive than buildings, despite the economies of mass production which are associated with their manufacture. The cost per tonne of steel used in a motor car is an order of magnitude greater than that of a structural steel framework for a building. The same is true of timber used for yacht construction

27

compared to structural timber in buildings and of the aluminium alloys used in aircraft construction. The high relative costs of the motor car, the sailing yacht and the aeroplane are not due to the basic costs of the materials but to the costs of fashioning these materials into complex shapes and configurations. By comparison with a vehicle, of whatever type, most buildings are very simple objects, being composed of basic components assembled into relatively simple geometries.

The virtually unlimited range of forms which is available to the designer of the motor car or the ship or the aeroplane is therefore at the disposal of the architect. The factors which have mitigated against the use of complex forms have for the most part, as stated above, not been technical and therefore not *real*, in the sense of ultimate physical, constraints.

There is, however, one technical factor which, in the context of buildings, does inhibit the freedom to invent form. This is scale, or more particularly the size of the span involved. If this is small, then the architect can indeed assume virtually unlimited freedom in the matter of form. The larger the span the more restrictive are the structural constraints on form.

The scale of a building is therefore of critical importance if its form is to be determined free from any consideration of structural performance. This is because the building will have to contain a structure with sufficient strength to resist the loads applied to it. If the form has been determined 'freely' in the visual sense (i.e. without consideration of technical implications) it is likely that the structure will be subjected to bending-type internal forces which will result in an inefficient use of structural material.[5] It is probable also that the magnitudes of these internal forces, for a given application of load, will be high, which will further compromise efficiency. If the internal

forces are so high that, even with strong materials such as steel or reinforced concrete, the sizes of the cross-sections required to contain stresses within acceptable limits are excessive, the structure will not be feasible. As is stated above the critical factor here is scale, and more particularly span.

This situation may be appreciated by considering the effect of size on any physical object. If the object is small – say less than one metre across (a model of a building for example) – it would be possible to make it from a weak material such as cardboard or papier mâché. If the span is 10 m (small building scale) a stronger material (reinforced concrete, steel, timber, plastic) would be required. The limits of what was possible would depend on the material and on the nature of the form. With reinforced concrete, for example, the limit could be reached at a span as low as 20 m but might not be reached until the span was 200 m, depending on the form. If steel were used the limits would be higher.

Scale is a significant consideration with any type of structure but it is particularly important if the relationship between structure and architecture is that of 'structure ignored'. The larger the span the greater is the possibility that the form will not be feasible. With the materials currently available, virtually any shape is possible up to a span of around 30 m. Most shapes are possible up to a span of around 60 m. For spans greater than 60 m there is an increasing possibility that a shape which has been chosen without regard to structural considerations will not be feasible. Built forms which have been determined without regard to structural considerations are likely to be significantly more costly to produce, however, than those which have.

To summarise, the architect who wishes to disregard structural considerations when determining the form of a building must be mindful of the considerations of cost and scale. If any significant departure is made from the basic forms of structure outlined in Chapters 3 to 6 here, then a cost penalty is likely to be involved. If the span is large there is also a possibility that the form chosen will not be

---

5 It will, in other words, most likely be either a semi-form-active or a non-form-active structure, neither of which is structurally efficient. See Appendix 1 for the explanation of this.

feasible. It would be imprudent, therefore, for an architect to adopt the 'structure ignored' methodology in the context of large enclosures, long spans and/or a parsimonious client.

2.2.3 Structure accepted

When this type of relationship between structure and architecture is preferred the objective is to produce a building in which more-or-less equal importance is attached to all aspects of the design. The aesthetic and programmatic issues must, as always, be brought to a successful conclusion but the structure must also be considered to be satisfactory when judged by purely technical criteria. It must not, in other words, be regarded simply as a provider of support which occupies the hidden, nether regions of a building and which does not, of itself, have to be well designed.

Buildings in which this type of relationship between structure and architecture exists have been constructed throughout the entire history of western architecture and include some of the most notable buildings in that tradition (Figs 5.4 and 5.7). The temples of Greek antiquity, the massive basilicas and bath houses of the Roman period and the Gothic cathedrals of medieval Europe are prominent examples. These have many equivalents in the modern world: the buildings by Le Corbusier which are based on reinforced concrete frameworks, for example (e.g. the Villa Savoye (Fig. 4.6), the Unité d'Habitation and the monastery of La Tourette), whose rectilinear geometries are ideally suited to economical forms of construction in that medium, are the twentieth-century equivalents of the Greek temples or the Gothic cathedrals, in terms of the relationship which was achieved between structure and architecture. The Willis, Faber and Dumas building by Norman Foster (Fig. 4.17) is a more recent example. The majority of buildings constructed in the present day, though not necessarily the most famous, also fall into this category.

In all of these buildings the structural and aesthetic programmes co-exist in harmony. The structures have not been expressed in an overt way but their properties, requirements and limitations have been accepted and visual vocabularies have been adopted within which they have been easily accommodated. The aesthetic and technical issues have, in other words, been reconciled.

To achieve this type of integrated design architects and structural engineers have to work together from an early stage in the design process. For a given size of building, the range of structural options which are sensible may be relatively small and an outcome in which all aspects of the design are deemed to have been satisfactorily resolved will be obtained only if a flexible strategy is adopted with regard to the building's appearance. The aesthetic strategy must, in other words, be compatible with the range of structural options which is most practical, given the spans involved.

The criteria by which the quality of a structure can be judged have been discussed elsewhere,[6] where it was argued that the primary technical objective of structural design was the satisfaction of load-carrying requirements with maximum economy of means. Well-designed structures were shown to be a compromise between complexity, which is required to economise on material, and simplicity, which allows economy to be achieved in the activities of design and construction. The principal factors which determine the optimum compromise between complexity and simplicity were shown to be span and the intensity of applied load. The larger the span, the greater was the level of complexity which was justified, because high levels of structural efficiency are required to achieve long spans. The greater the applied load, the lower was the level of structural efficiency which could be tolerated and the simpler therefore the structural form which was appropriate.

Thus, it was seen that simple structural forms composed of basic elements (timber beams or reinforced concrete beams and slabs with simply shaped cross-sections) assembled into basic post-and-beam arrangements are appropriate for short-span structures. These

---

6 See Macdonald, *Structure and Architecture*, Chapter 6.

represent the best compromise between complexity and simplicity for buildings which, though not necessarily of small scale, have no requirement for large interior spaces and therefore for long spans. At the other extreme – the very long span – highly complex systems such as steel-cable networks or doubly-curved reinforced concrete shells were seen to provide the best structural solutions. For intermediate spans, intermediate levels of complexity are appropriate.

A happy consequence of regarding the achievement of maximum economy of means as the principal criterion of good structural design is that it is also the condition which is likely to result in the lowest-cost structure. Although monetary cost is, in important ways, an artificial yardstick, it is in fact related to the totality of the resources, of all kinds, which must be committed to a structure.

A factor which affects the precise level of complexity which is appropriate for a particular structure is the economic climate in which it will be built. If, for example, the cost of labour is low in relation to that of materials, which is the normal situation in a non-industrialised economy, a more complex (and therefore more efficient) structural form is justified than if the reverse is true. The structure type which is most appropriate for a particular span in an industrialised economy, in which labour costs are high, might therefore be simpler than the most appropriate type of structure for the same span in a 'developing' country.

As was noted above, good examples of well-resolved architecture of the 'structure accepted' type are found in the early modern period. This was due to the happy compatibility of the favoured orthogonal aesthetic with the post-and-beam structure, which is the most appropriate structure type for the relatively short spans which occur in the majority of buildings.[7]

To summarise, the building designer who wishes to achieve technical as well as aesthetic and programmatic excellence must select a type of structure which is truly appropriate for the span and load conditions involved. This is likely to have a fairly profound effect on the overall form of the building and in such a case the technical and aesthetic agendas must be made compatible. Because the range of structural options may be limited, particularly by the span involved, the approach to the aesthetics must be flexible. Where this methodology is adopted the structural solution will inevitably be an adaptation of one of the basic forms of structure outlined in the following chapters.

2.2.4 Structure symbolised

One of the features of buildings designed in accordance with the 'structure accepted' approach discussed above is that the structure itself rarely constitutes a prominent part of the visual vocabulary and may even be entirely hidden from view. The objective of the 'structure accepted' approach is not overly to express the structure visually but to ensure that it will be well designed and well integrated with all other aspects of the building. Where the relationship between structure and architecture is of the 'structure symbolised' type, the structure is emphasised visually and constitutes an essential element of the architectural vocabulary.

In the 'structure symbolised' approach the structure is treated as a set of visual motifs and decisions concerning the size, shape and arrangement of the structural elements are influenced as much by visual as by technical criteria. The technical performance of the structure is secondary to its aesthetic role and the technical quality of the structural design is frequently compromised as a result. This is an inevitable consequence of this method of working and must be accepted if it is used.

In recent architecture the 'structure symbolised' approach has been employed almost exclusively as a means of expressing the idea of technical progress and has therefore been associated principally with the Modern

---

7 In the context of architectural structures the post–and–beam arrangement is non–form–active. For short spans, this is, in most cases, the best *structural* solution. See Macdonald, *Structure and Architecture*, Chapter 6 for a discussion of this.

Fig. 2.6 Renault Warehouse and Distribution Centre, Swindon, England, 1983. Foster Associates, architects, Ove Arup & Partners, structural engineers. Many of the features which are associated with high structural efficiency are visible here, such as the semi-trussing of the main elements, the use of tapered profiles, the I-shaped cross-sections and the circular lightening holes. They were adopted for visual reasons, however, as few of these complexities can be justified on technical grounds due the relatively short span involved. [Photo: Alastair Hunter]

Movement. In the early Modern period, in the works of architects such as Mies van der Rohe, the use of technical imagery was accomplished with restraint and refinement, and confined to the repetitive use of elements such as parallel-sided I-section beams (see Fig. 1.2). Later practitioners of the so-called 'high-tech' genre used a much richer palette of structural images, many of which were 'borrowed' from the fields of vehicle and aeronautical engineering rather than from structural engineering. The purpose was to create an architectural style which was celebrative of the *idea* of technical progress and of the condition of modernity.

In the high-tech style the 'structure symbolised' approach resulted, with a few notable exceptions, in an architecture which was based on steel framework structures, principally because steel was the only structural material which possessed the necessary high-tech image and which also provided a large enough number of different element and component types to give the architectural language a reasonably large vocabulary.[8]

The use of steel frameworks in this way meant that the constraints which are associated with steel had to be accepted and the

---

8 It is significant that one of the very few buildings of this type to have a reinforced concrete structure – the Lloyd's Headquarters Building in London – was intended originally to have a steel frame and was detailed as though the structural material were indeed steel. This provides a good illustration of the fact that, contrary to the claims which are often made for it, this is not a structurally 'honest' type of architecture.

building forms which resulted therefore had to have rectanguloid geometries with regular layouts of elements. The shapes of the individual structural elements were more complex, however, than would have been specified for equivalent structures under the 'structure accepted' philosophy and the various devices which are associated with high structural efficiency were freely employed to add visual interest. These included complex shapes in cross-section and longitudinal profile, the tapering of elements, the liberal use of the circular lightening hole (used in aircraft structures to reduce the weights of components) and the exaggeration of joints and fastening components (Fig. 2.6). Much of this vocabulary was 'borrowed' from aerospace or motor technology, where its use was, of course, justified on purely technical grounds. In the field of architecture, its employment was purely stylistic but was necessary to convey the impression of 'state-of-the-art' technology.

The need to expose the structure to view, which is an inevitable consequence of its use as a set of visual motifs, carried with it a requirement to provide adequate corrosion protection schemes, which added to the structural costs, especially if, as was often the case, the steelwork was exposed on the exterior as well as in the interior of the building. Where the building had more than one storey, it was also necessary to make adequate provision for fire protection. This posed quite severe problems which were often difficult to resolve[9] and which frequently required the adoption of one of the more complex forms of fire protection (see Section 3.5).

One of the ironies of the high-tech movement was that when structural form was manipulated according to the dictates of aesthetic rather than technical criteria, the result was often technically flawed and it is a fact that most of the high-tech buildings were not well engineered if judged purely on their technical performance.[10] That which symbolised technical excellence was not, in other words, itself technically excellent. This, however, was an inevitable consequence of the symbolic use of structure.

It should be noted, however, that the 'structure symbolised' approach is capable of being applied to architectural agendas which are different from the Modernist preoccupation with the celebration of technology. If, for example, the intention was to celebrate the idea of a sustainable architecture the use of the 'structure symbolised' method would yield a quite different architecture in which attention was drawn to the measures which are required to minimise the input of energy and resources, of all kinds, which are involved in the construction, maintenance and running of a building. Such an approach could lead to the development of a completely new architectural symbolic vocabulary of masonry and timber. This would, of course, also be equally 'dishonest' and probably as basically inefficient, in structural terms, as was 'high tech'. This development is therefore perhaps less likely to occur than was 'high tech' because the potential clients would, due to its lack of engagement with physical reality, be less sympathetic to the idea of adopting it and more likely to favour the 'structure accepted' or 'true structural high-tech' approach.

Where the 'structure symbolised' approach is used it is essential that the designer has a clear awareness of the type of architectural statement which the building is intended to make. The form of the structure must then be distorted and exaggerated so as to make the nature of that statement unequivocally clear otherwise the result will be architecturally feeble.

2.2.5 True structural high tech
In this, the fourth type of relationship between structure and architecture, the design of the structure is accorded the highest priority and

---

9 It was in fact due to the impossibility of meeting the fire regulations that the change of structural material from steel to reinforced concrete was made in the case of the Lloyd's building.

10 See Charles Jencks, 'The battle of high tech: great buildings with great faults', *Architectural Design*, **58**, (11/12) pp. 19–39, 1988.

allowed to determine completely both the overall form of a building and the nature of the architectural vocabulary which is adopted. It is a method which is normally used for reasons of necessity, when the limits of what is possible structurally are being approached. Obvious examples of this are the very tall building and the very long-span enclosure. In the case of the former, the principal structural problem is the resistance of lateral load, in which case the systems required to accommodate this become prominent features of the design.[11] The achievement of a long span requires that a highly efficient type of structure be adopted, such as a steel cable network or a reinforced concrete shell. These form-active[12] structure types have distinctive geometries which therefore dictate the overall form of the building.

Another example of a building type in which the highest priority must be given to structural matters is the portable building. Here there is a requirement for demountability and also a critical need to save weight and therefore to adopt an efficient form of structure. Form-active structures are therefore frequently specified for this type of building. The tent, in all its manifestations, provides an example of this. Demountable buildings for travelling exhibitions (Fig. 1.12) and any other type of temporary accommodation are further examples. With this type of building very few concessions are made to style, and none which conflict with efficiency.

The 'true structural high-tech' approach can be applied to the design of any building. Architects rarely favour it, however, and will rarely use it unless technical necessity makes it unavoidable, because it leaves them with little to do other than administrate the construction of the building. It leaves, in other words, little

scope for artistic statement. The methodology requires that no concessions be made with regard to the design of the structure and the resulting architecture might be described as a type of vernacular. For the majority of buildings, in which spans and loads are modest and there are no special requirements which favour the adoption of a highly efficient form of structure, the best structural solution is likely to involve the use of a very simple type of structure such as a post-and-beam form composed of basic elements such as masonry walls or reinforced concrete slabs and columns.

The 'true structural high-tech' approach is the most straightforward of the possible relationships between structure and architecture. The preliminary design of the building becomes simply the design of a structural arrangement which is appropriate for the span and load involved. In most cases this will favour the selection and adaptation of one of the basic forms of structure outlined in the following chapters. Aesthetic, space-planning and other considerations are given a secondary role and are not allowed to compromise the integrity of the structural solution.

2.2.6 Conclusion

The various possibilities concerning the relationship between structural design and architectural design have been reviewed in this section. The distinctions considered have only been possible in the twentieth century following the development of the structural technologies of steel and reinforced concrete – the strong materials which released architecture from the constraints imposed by the technology of structural masonry. It was not until this had occurred that the methodologies of 'structure ignored' and 'structure symbolised', which have a tendency to generate structural geometries which are far from ideal, became possible.

The distinctions between the four approaches outlined above to the relationship between structure and architecture are frequently misunderstood, by both architects

---

11 In all of the very tall buildings the structure which is required to resist wind loading is concentrated on the exterior of the building and therefore affects its appearance.

12 See Appendix 1 for an explanation of the term 'form–active'.

and critics, especially in connection with those strands of Modern architecture in which the idea of drawing attention to the tectonic aspects of a building has been fashionable. The confusion which arises concerns the allocation of priorities between technical and non-technical issues. If technical issues have, in reality, been allocated the highest priority the building will fall into the category of 'true structural high tech'. If aesthetic considerations have been given a higher importance then the building will be an example of 'structure symbolised' or 'structure ignored'.

The distinction is brought into focus by a consideration of the consequences of allowing a high design priority to be given to technical considerations ('structure accepted' or 'structural high tech'). If the structural problem is spectacular, such as a very long span, the resulting structure, and therefore building, will also be visually striking. If the structural problem is modest – a building of small or medium span – the best structural solution will almost certainly also be modest and of the post-and-beam kind. In the early years of architectural Modernism – the 1920s and 1930s – such forms were compatible with the prevailing aesthetic theories and, as a consequence, many early modern buildings are good examples of the 'structure accepted' approach. In the present day, post-and-beam forms are frequently considered to be visually dull and in this situation the temptation arises to manipulate the structure for visual or symbolic reasons ('structure symbolised') or to ignore its requirements entirely ('structure ignored').

As is discussed above, the architectural symbolists of the so-called 'high-tech' school, who have often claimed that the structures of their buildings are examples of genuine technical excellence,[13] provide a good example of the type of unclear thinking which has surrounded this topic. The confusion has led, in many cases, to the creation of buildings which have

an unresolved quality, because the full potential offered by the purely symbolic use of structure has not been exploited. A better architecture would probably have resulted if the true nature of the relationship between structure and architecture had been more fully appreciated and acknowledged.

This last statement is generally true: the final outcome of an architectural design process is more likely to be satisfactory if the architect is fully aware of the nature of the relationship between technical and aesthetic issues.

2.3 Selection of the generic type of structure

2.3.1 Introduction

Structures for buildings may be placed into the three broad categories of 'form-active', 'semi-form-active' and 'non-form-active'.[14] In the context of gravitational loading, which is the principal form of load on most architectural structures, post-and-beam structures are non-form-active and can be further subdivided into the two categories of loadbearing-wall structures and skeleton-frame structures. It is from this limited range of possibilities that the structure for a building must be selected. Within each category an almost infinite variety of structural possibilities exists, however, depending on the types of element which are specified and the manner in which these are connected together.

It is important to recognise that the process of structural design is not so much one of *invention* as one of *selection and adaptation*. Most new structures are in fact versions of one of a range of basic structural forms which have evolved in practice as the best arrangements for the particular material concerned. These basic forms of structure are reviewed in Chapters 3 to 6 and they represent the vocabulary of structural form from which the designer

---

13 The late modern deconstruction architects have not been troubled by this particular item of theoretical baggage.

14 See Appendix 1 and Macdonald, *Structure and Architecture*, Chapters 4 and 5.

must begin the process of selection and adaptation. An insight into the working of this process is given in Section 2.5.

An important factor in the selection of a structure type for a building is the nature of the relationship which has been adopted between structure and architecture. If this is in the 'structure ignored' category, then the overall form of the building, and therefore of the structure, will have been determined without consideration of structural require-ments and the structure will most probably have to be of the semi-form-active type. This is because, for a given load condition, form-active and non-form-active structures have unique geometries. A structure whose geom-etry has been determined without regard to structural considerations is therefore likely to have a semi-form-active shape. In this situ-ation the choices available to the structural designer may be very limited.

The structural material will almost certainly have to be either reinforced concrete or steel because a material with the ability to resist significant bending will be required. The choice between these two materials will be influenced by the nature of the building. Concrete results in a more massive structure but allows the easy creation of curvilinear forms. Steel is stronger and is therefore capable of producing a lighter structure (both literally and in appearance) but must normally be used in the form of a skele-ton framework. The distinction between their respective suitabilities can be seen by compar-ing the Vitra Design Museum building by Frank Gehry (Fig. 4.20) with the rooftop office in Vienna by Coop Himmelblau (Fig. 2.1).

In the Vitra Design Museum building the presence of strongly curvilinear forms, of cantilevered volumes of complex shape and of the limited number of openings in the build-ing's envelope favoured the use of reinforced concrete. The mouldability of the material and the level of structural continuity which it allows are particularly useful properties. In the Rooftop Office in Vienna the extensive areas of faceted glazing favoured the adoption of a skeletal structure of slender steel elements. The minimal amount of curvature in the roof

structure was accommodated by steel and the general arrangement of the structure, despite its apparent novelty, conforms to the fairly standard primary/secondary beam format.

If the relationship between structure and architecture is other than 'structure ignored' it will be necessary to select a structure type which is compatible with the aesthetic and programmatic aspects of the building and which is also sensible from a structural point of view. The choice of structure type should, however, be larger because there should be a readiness to adjust the aesthetic programme to accommodate structural requirements and thus produce a building with a structure which is satisfactory technically. The factors on which this depends are now reviewed.

2.3.2 The effect of scale

The span of an architectural structure, which is determined by the required sizes of the spaces which are enclosed by it, has a very significant effect on both the generic type of structure which should be used and on the selection of the types of structural element of which it should be composed. The underlying principle which governs the relationship between span and structure type is that the ratio of self-weight to load carried should be satisfactory. For a given type of structure the strength-to-weight ratio tends to become less favourable as the span increases.[15] More efficient types of structure must therefore be specified as spans are increased to maintain the ratio at an acceptable level.

Table 2.1 shows the ranges of span for which basic types of structure are most suitable and therefore normally specified. The figures must be regarded as approximate but serve to give an indication of the applications for which each is most appropriate. Separate figures are provided for floor and roof structures to allow for the variations which are caused by the significantly different levels of gravitational load to which they are subjected.

15 See Macdonald, *Structure and Architecture*, Chapter 6.

Table 2.1   Normal span ranges for commonly used structural systems

A Timber structures

Structural system	Normal span range (m)		Span/depth ratio
softwood planks	floor	0.6–0.8	25–35
	roof	2–6	45–60
plywood board	floor	0.3–0.9	30–40
	roof	0.3–1.2	50–70
softwood timber joist	floor	3.5–6	12–20
	roof	2–6	20–25
stressed skin timber panel	floor	3–6	20–30
	roof	3–9	30–35
laminated timber beam	floor	5–12	14–18
	roof	4–30	15–20
plyweb beam	floor	5–18	8–10
	roof	6–20	10–15
trussed rafter	roof	5–11	4–6
parallel chord truss	roof	12–25	8–10
gluelam portal frame	roof	12–35	30–50
gluelam arch	roof	15–100	30–50
lattice dome	roof	15–200	40–50

B Steel structures

Structural system	Normal span range (m)		Span/depth ratio
profiled decking	floor	2–3	35–40
	roof	2–6	40–70
profiled decking with composite concrete topping	floor	2–6	25–30
beam (hot-rolled section)	floor	6–25	15–20
	roof	6–60	18–26
cold-formed open-web joist	roof	4–30	15–25
hot-rolled triangulated parallel-chord truss	floor	12–45	4–12
	roof	12–75	10–18
pitched, triangulated truss	roof	25–65	5–10
space deck	roof	10–150	15–30
braced barrel vault	roof	20–100	55–60
portal framework	roof	9–60	35–40
cable-stayed roof	roof	60–150	5–10
hanging cable roof	roof	50–180	8–15

C Reinforced concrete structures

Structural system	Normal span range (m)		Span/depth ratio
one-way span solid slab	reinforced	2–7	22–32
	prestressed	5–9	38–45
two-way span solid slab	reinforced	4.5–6	30–35
one-way span ribbed slab	reinforced	4–11	18–26
	prestressed	10–18	30–38
two-way span coffered slab	reinforced	6–15	18–25
	prestressed	10–22	25–32
beam/column frame with beam/slab floor	slab	3.5–6	30–36
	beam	6–12	15–20
portal framework		12–24	22–30
arch		15–60	28–40

All structure types have a potential maximum span and the less efficient the structure, the lower is this practical maximum span. An indication of this is given in Table 2.1 in which it will be observed that the maximum span given for each type of structure is related to its potential efficiency. Thus, elements with simple, solid rectangular cross-sections, such as sawn timber joists or rectangular cross-section reinforced concrete slabs, have relatively low maximum spans. Simple elements with 'improvements', such as I-section steel beams or triangulated trusses of timber or steel, have higher maximum spans. The highest spans are achieved by highly efficient vaulted shells or cable networks.

The effect of the variation in the maximum span potential of different types of element is that the choice of element types which is available to the structural designer diminishes as the span increases. If the span to be achieved is small (say 5 m) virtually all structure types are available. At this scale the designer could therefore choose to use any type of structure from simple, solid beams or slabs to sophisticated forms such as the arch or the vault. In the context of contemporary architecture, it would probably be regarded as technically inappropriate to use a complex form for a short span, unless some special requirement for high efficiency existed, because much simpler post-and-beam forms would perform adequately. It would nevertheless be a choice which the architect could make. As the span increases the number of different types of structure which would be viable decreases until, at the very long span (say 200 m), only the most efficient form-active types, such as steel cable networks or thin concrete shells, are feasible. In summary, from the point of view of the designer, the choice of structure type is large if the span is small and becomes progressively more limited as the span increases.

The most basic post-and-beam types of structure (loadbearing-wall arrangements in masonry or timber with simple horizontal elements such as timber beams or reinforced concrete slabs) are suitable for short-span structures in the 5 m to 10 m range. The span range can be extended by the use of more efficient types of horizontal structure such as triangulated trusses in either timber or steel. The use of walls with 'improved' cross-sections (fin wall or diaphragm wall) allows the very basic structural system to be used for larger enclosures with high external walls, such as sports halls, where spans of up to 30 m have been achieved.

The post-and-beam frame (in either reinforced concrete or steel) is a more sophisticated and therefore more flexible system than the loadbearing wall. In the multi-storey version the span range is slightly more extensive than that for loadbearing-wall structures (5 m to 20 m). The most basic types of element are used at the low end of the range (solid reinforced concrete slabs, rolled-steel sections) and at the upper end more efficient types such as coffered, reinforced concrete slabs and hollow-web steel beams. Spans greater than 20 m are unusual in multi-storey buildings but where these occur, efficient types of structural elements must be specified (e.g. triangulated girders used to achieve a span of around 25 m at the Centre Pompidou in Paris).

In single-storey structures the change from the most basic forms to more efficient structure types (e.g. space frame horizontal structures or semi-form-active structures) normally occurs when the span is in the range 20 m to 30 m, with spans greater than 30 m usually requiring the use of a semi-form-active structure such as a portal framework. As with short-span structures, the type of element which is used can vary and the tendency is always for the level of efficiency (and therefore of complexity) to increase as the span increases. Thus, short-span portal frameworks (20 m) would normally be accomplished with rolled-steel sections such as the universal beam while a triangulated-truss longitudinal profile might be specified for a longer-span version (say 50 m).

The transition from the semi-form-active to the highly efficient fully form-active structure occurs in the range 40 m to 60 m.

2.3.3 The effect of cost

Another factor which influences the choice of structure type for a particular span is cost. Structural efficiency is achieved by structural complexity which is costly to produce. The builder of the long-span structure has no choice but to use an expensive, complex type of structure. The builder of the short-span structure does have a choice: both complex and simple types will be feasible, but the complex structure will be more expensive. The use of the latter, in the cost-conscious environ- ment of the late twentieth century, would therefore normally have to be justified on some other grounds such as a particular requirement for a lightweight structure. The reason for choosing a complex form might, of course, be simply a desire for the appearance of complexity or extravagance. Depending on the priorities of the client, therefore, cost might or might not be a significant factor in determining the structural form which is adopted.

2.3.4 Internal planning

The nature of the plan which is intended for a building will normally influence the choice of structure. The principal factors which affect this are the degree to which the interior is subdivided, the degree of regularity required in the internal planning and the degree of en- closure which is envisaged.

The extent to which a building will be compartmentalised normally has a consider- able influence on the choice of structure type. Multi-cellular buildings lend themselves to loadbearing-wall-type structures, subject to the provisos that the individual compartments are not too large, that a reasonably regular pattern of walls can be adopted and, in multi-storey buildings, that the plan is more or less the same at all levels. If any of these conditions cannot be met, then a frame structure will normally be required even if the plan is compartmentalised.

If the interior of a building is to consist of large areas of open space a frame structure will normally be required even if the space can be interrupted by some vertical structure – for

example by columns. There have, of course, been exceptions to this, the 'prairie houses' of Frank Lloyd Wright being good examples of interiors with free flowing internal spaces being accommodated within loadbearing-wall structures. These buildings are of relatively small scale, however, and the free-flowing spaces exist within patterns of walls which are fairly regular (Fig. 2.7).

The degree of regularity of a plan affects the preferred type of structure. As is mentioned above, loadbearing-wall structures must normally be given a regular arrangement (see Chapter 5) and the same is true of frames, which perform best if based on a regular column grid. If the pattern of vertical support is not regular, this is best accommodated by the adoption of a system for the horizontal structure which has the capability to span in more than one direction. For floor structures the simplest element of this type is the two- way-spanning slab of *in-situ* reinforced concrete. For roof structures it is the fully triangulated space deck.

Although both of these structural types work best when provided with a regular pattern of vertical support they can, due to their statical indeterminacy,[16] be supported on irregular arrangements of columns or loadbearing walls. The greater the degree of irregularity the stronger, and therefore deeper, must these elements be.

If the irregularity of the design of a building extends to a variation in plan between different floor levels, this is best accommodated by the adoption of a frame-type structure. The walls which subdivide the interior are then non- loadbearing partitions which are simply carried by the floors and whose location in the plan can be varied between levels. If the variation in plan between levels is such that the continuity of columns through different levels is not

---

16 See Macdonald, *Structure and Architecture*, Appendix 1 for a discussion of statical indeterminacy in relation to structural design.

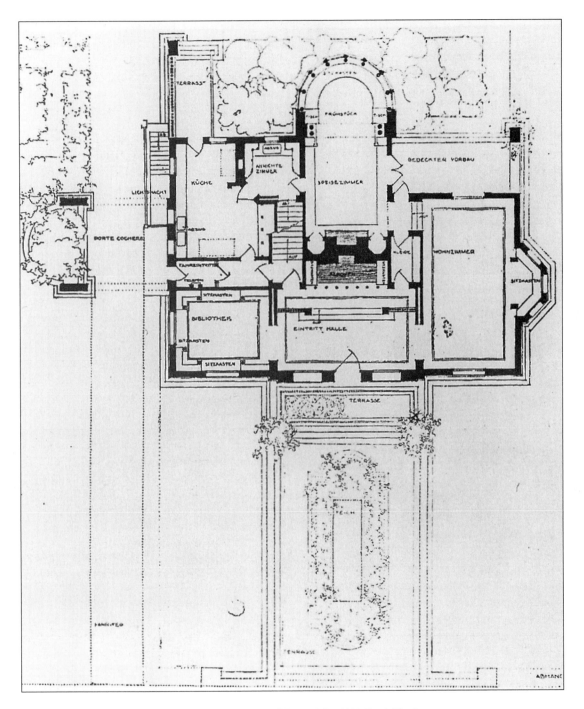

Fig. 2.7 Plan of the Winslow House, River Forest, near Chicago, USA, 1893. Frank Lloyd Wright, architect. The open plan of the Winslow House is accommodated within a conventional arrangement of parallel loadbearing walls (see Fig 5.23) with lintel beams spanning the large openings between the principal spaces. The small scale of the building together with the relatively large areas of vertical structure which are provided have allowed the open interior to be achieved without the use of an expensive framework structure.

possible, this must be achieved by the use of substantial horizontal structure at the levels at which discontinuities in vertical structure occur. A deep structure will be required, however, and appropriate provision for this must be made.

If the form of a building is highly irregular in plan, elevation or section an unconventional form of structure will be required. This will normally have to be based on either reinforced concrete or steel – see above.

2.3.5 Exterior treatment

The treatment which is envisaged for the exterior of a building can affect the choice of structure type, especially if large areas of the walls will either be glazed or covered with a cladding material which is incapable of acting as part of the support structure. Full glazing of the exterior walls, in particular, is normally associated with a skeleton-frame structure. It is worth noting, however, that this type of treatment of the exterior need not, by itself, necessarily dictate that a frame structure be adopted as it is possible to arrange the plan of a loadbearing-wall building in such a way that the exterior walls are non-loadbearing (e.g. with a cross-wall-type plan). Full glazing of the exterior of a building is, however, normally associated with a relatively open plan for which a frame structure would in any case be appropriate.

2.3.6 Conclusion

The selection of the generic type of structure is therefore based on a number of factors, the most important of which are: scale, which determines the spans involved; internal planning which dictates the nature of the internal spaces required; and external treatment. The structural choices available are form-active, semi-form-active and non-form-active structures, in all their manifestations. The final selection will be determined by the issues of cost and technical feasibility, given the requirements of scale, space planning and external treatment.

2.4 Selection of the structural material

The choice of the structural material is another fundamental decision in the planning of a structure. It is an aesthetic as well as a technical decision. Each of the four principal structural materials (steel, reinforced concrete, masonry and timber) produces a building with a distinctive visual quality.

The preferred generic type of structure will also affect the choice of structural material and indeed these two issues are frequently resolved together. If, for example, a loadbearing-wall structure is adopted this will favour the use of either masonry or timber. Reinforced concrete and steel would be the normal choices for skeleton-frame structures.

If a loadbearing-wall structure is being used for a domestic-scale building there may be little to choose between masonry or timber from the point of view of structural action. A masonry wall will normally have greater strength in relation to compressive load,[17] which means that the use of masonry is essential if the building has more than three storeys.[18] It also favours the use of masonry in situations where internal spaces are large (greater than 8 m across) and in other situations, such as the adoption of an irregular pattern of loadbearing walls, in which the loads applied to individual walls are likely to be high. Other considerations, such as local availability of materials, speed of erection or any requirement for the prefabrication of the

---

17 Not because the material is inherently stronger but because the volume of the structure itself, for a given thickness of wall, is greater. The timber wall will consist of a series of closely spaced vertical elements alternating with voids. The masonry wall is continuous and therefore contains more structural material for a given thickness.

18 Traditional half–timbered buildings were frequently higher (six or seven storeys), especially in the towns of northern Europe. These were based on very large timber sections, however. This type of structure is very rare in the present day.

structure are, however, more likely to influence the choice of structural material than the strength required.

Skeleton-frame structures can be constructed in steel, reinforced concrete or timber. The use of timber is rare for multi-storey buildings, however, and the timber skeleton frame is most usually associated with single-storey enclosures in which the timber structure is exposed to view. Often the structural elements are of spectacular appearance (large triangulated trusses) and the structure therefore contributes significantly to the appearance of the building. The relationship between structure and architecture must normally be of the 'structure accepted' or 'true structural high-tech' type rather than the 'structure symbolised' type, when timber is used for skeleton frames. This is because there is normally insufficient reserve of strength to allow the structural performance to be compromised for purely visual effect.

Reinforced concrete is rarely used for single-storey buildings because the high self-weight of the structural elements renders them unsuitable for situations in which the imposed loads are small. For the same reason, reinforced concrete is rarely used for roof structures in multi-storey buildings except for flat roofs, in the form of terraces or roof gardens, to which access is permitted and which are therefore required to carry greater than normal roof loadings. Steel is suitable for both multi-storey and single-storey buildings.

The normal choices available to the designer of a skeleton-frame structure are therefore between steel and timber for single-storey frames and between steel and reinforced concrete for multi-storey frames. The relative advantages and disadvantages of the materials, from a structural point of view, are outlined in Chapters 3, 4 and 6.

The normal reasons for the selection of steel are its high strength, its appearance (especially in the context of the 'structure symbolised' approach) or the speed of construction which it allows. The first of these is obviously important if long spans are involved or if there is a need to produce a structure which is of low volume and which has slender elements. Appearance is frequently also the principal reason for the selection of timber for a skeleton framework although the lightness of the elements can also be a factor. Timber may also be selected due to its durability, for example where hostile internal environments, such as occur in swimming pool buildings, would cause corrosion of a steel structure.

Reinforced concrete can be selected for a number of reasons. Often, it will provide the lowest cost structure due to both the low inherent cost of the material and the elimination of the need for additional finishing materials or fire proofing. The good durability of concrete is one of its considerable advantages. Concrete may also be selected due to the opportunities which it provides in relation to form. The mouldability of the material and the ease with which it allows structural continuity to be achieved favour its use in situations in which complex structural geometries are to be produced, especially if these involve curvilinear forms.

All materials impose constraints as well as provide opportunities. Thus, unless the 'structure ignored' method is being practised, the limitations of the material must normally be recognised once it has been chosen. The nature of these constraints is something which is taken into account during the selection of the material to ensure that it is compatible with the architectural intentions.

2.5 Determination of the form of the structure

The final form and general arrangement of a structure is normally an adaptation of one of the basic forms described in Chapters 3, 4, 5 and 6. This is now discussed in relation to reinforced concrete structures to illustrate how the process of adaptation is carried out to produce buildings in which structural and architectural requirements are reconciled. The methodology described is applicable to all structural materials.

Fig. 2.8 Plan of the Willis, Faber and Dumas office, Ipswich, England, 1974, Foster Associates, architects, Anthony Hunt Associates, structural engineers. The floors of this building are reinforced concrete coffered slabs which are supported on a conventional square column grid to give optimum two-way-spanning capability. Small-diameter columns at close spacings are provided around the perimeter to support those parts of the plan which project beyond the main column grid. The complex curvilinear plan-form has been achieved by a relatively minor adaptation of the basic arrangement for two-way-spanning slabs (see Fig 4.47).

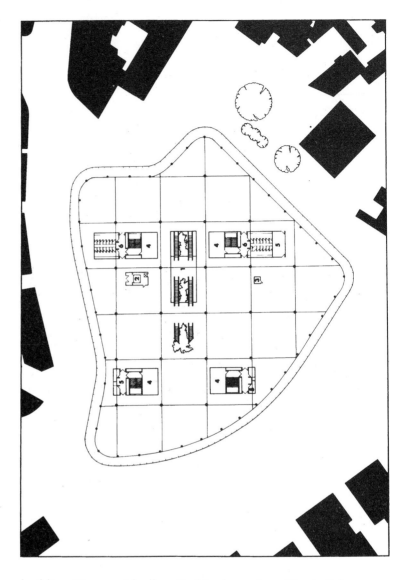

The Willis, Faber and Dumas building (Fig. 4.17) is a good example of a two-way-spanning flat slab structure of reinforced concrete. The building has three storeys and a very deep plan consisting mostly of open-plan office accommodation. The requirement for the latter ruled out the possibility of using a loadbearing-wall structure. The curvilinear plan of the building favoured the use of reinforced concrete rather than steel as the structural material.

The floor slabs are supported on a square column grid which conforms to the standard plan for this type of two-way-spanning structure (Fig. 4.47). It is a configuration which is ideally suited to deep-plan office buildings. The flat-slab structure depends for strength on structural continuity and performs best in situations in which several structural bays are provided in two orthogonal directions. It is a form of structure which is suitable for relatively high levels of uniformly distributed imposed load as in the standard pattern of office loading. The absence of deep beams in flat-slab structures facilitates the creation, by use of a false ceiling, of a continuous services zone under the floor structure and the straightforward layout of reinforcement and relative simplicity of the structural form makes the

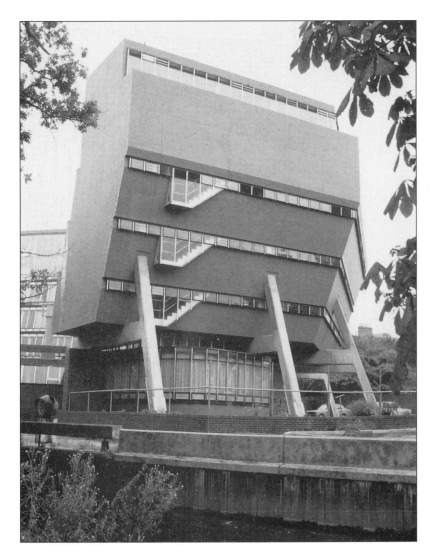

Fig. 2.9 Florey Building, Oxford, England, 1967–71. James Stirling, architect, Felix Samuely & Partners, structural engineers. This building has a fairly complex form which was realised by a relatively straightforward adaptation of a standard reinforced concrete framing system. [Photo: P Macdonald]

construction process simple and therefore inexpensive.

In the Willis, Faber and Dumas building the curvilinear plan-form conflicted with the square column grid and the problem was resolved by the use of a series of small-diameter perimeter columns (Fig. 2.8). This technique by which the incomplete squares of slab in the main structural grid were provided with the support which they required was made possible by the two-way-spanning capability of the floor slab. The relatively large span involved (16 m) justifies the use of the coffered version of the flat slab. The distinctive form of the Willis, Faber and Dumas building

was therefore made possible by a skilful adaptation of one of the basic forms of reinforced concrete structure.

The Florey Building in Oxford, England (Figs 2.9 and 2.10) has a reinforced concrete frame structure with a one-way-spanning floor deck. This is a university hall of residence with a multi-cellular interior and it is also a relatively small building. These factors would normally favour the use of a loadbearing-masonry structure but the distinctive cross-section of the building, together with the requirement for an open plan on the ground floor, resulted in the selection of a skeleton framework in this case. The relatively complex form in both plan and

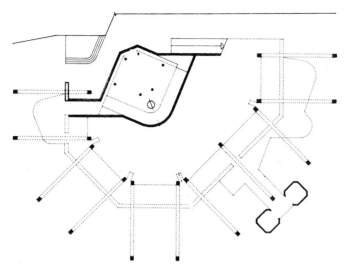

Fig. 2.10 Structural plan of the Florey Building, Oxford, England, 1971. James Stirling, architect, F.J.Samuely & Partners, structural engineers. The structural plan of this building shows the layout of the primary elements, which are beam/column frameworks of *in situ* reinforced concrete. These carry a one-way-spanning floor system. As in the case of the Willis, Faber and Dumas building a complex form has been achieved by the straightforward adaptation of a basic structural arrangement (see Fig 4.52).

Fig. 2.11 New City Library, Munster, Germany, 1993. Architekturburo Bolles-Wilson, architects, Buro Thomas, Buro Menke & Kohler, structural engineers. The photograph shows the edge of one of the two parts of this building. The main part of the building is to the right and has a reinforced concrete frame structure. One of the lines of vertical structure is seen and it will be noted that this is, in effect, a loadbearing wall which has been punctured in places by large openings to leave pillars of varying cross-sectional shape. The non-structural cladding wall seen on the left of the photograph is supported by a lean-to arrangement of cranked laminated timber elements. [Photo: E. & F. McLachlan]

cross-section favoured the use of reinforced concrete rather than steel.

The choice of a beam/column framework rather than a flat-slab structure was sensible for two reasons. Firstly, the depth of the building is that of a single structural bay. The structural benefit of continuity over several structural bays, which is desirable for the efficient working of a flat-slab structure, was therefore not available. The crescent-shaped plan was also problematic in this respect because it reduced the level of structural continuity which could be achieved in the along-building direction. Secondly, the cross-section of the building, and in particular the outward cant of the rear wall, favoured the use of rigid frames consisting of wide columns and deep beams. A beam/column framework in reinforced concrete was therefore the logical choice of structure for this building.

The crescent-shaped plan was produced by simply distorting the basic plan-form of the one-way-spanning frame (Fig. 4.52) to create a structural plan which was in fact an alternating series of triangles and rectangles. The arrangement, and in particular the triangular parts of the plan, were more easily constructed in reinforced concrete than would have been possible with steel. The mouldability of concrete was also exploited to create the distinctive cross-section of the building. Its durability allowed the elimination of additional finishing materials over most of the structural surfaces.

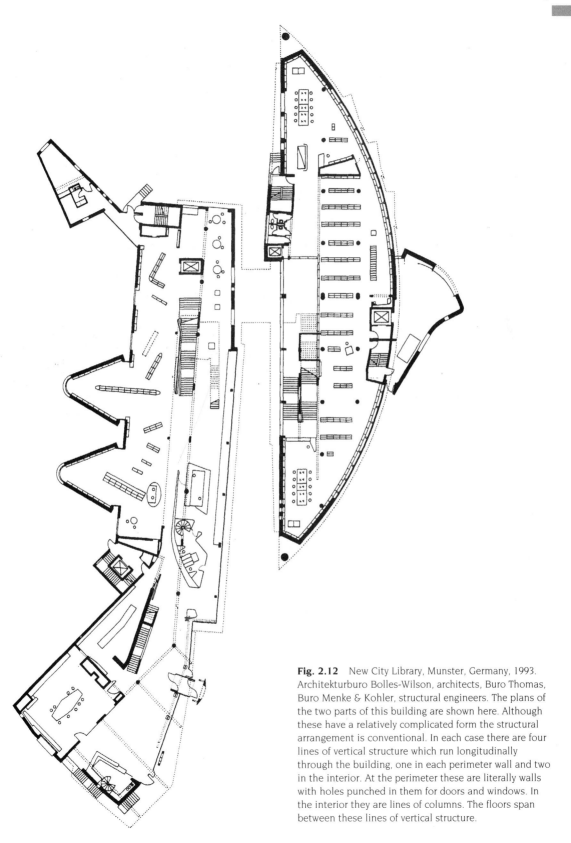

Fig. 2.12 New City Library, Munster, Germany, 1993. Architekturburo Bolles-Wilson, architects, Buro Thomas, Buro Menke & Kohler, structural engineers. The plans of the two parts of this building are shown here. Although these have a relatively complicated form the structural arrangement is conventional. In each case there are four lines of vertical structure which run longitudinally through the building, one in each perimeter wall and two in the interior. At the perimeter these are literally walls with holes punched in them for doors and windows. In the interior they are lines of columns. The floors span between these lines of vertical structure.

Fig. 2.13 Exhibition and Assembly Building, Ulm, Germany, 1993. Richard Meier, architect. The relatively complex form of this building was accommodated by a structure which was a relatively straightforward adaptation of one of the standard reinforced concrete arrangements. [Photo: E. & F. McLachlan]

The distinctive form of this building was therefore made possible by an imaginative exploitation of the structural possibilities of reinforced concrete which involved a relatively simple modification of one of the standard plan-forms for multi-storey reinforced concrete structures. The structure stands up well to purely technical criticism and the building falls into the category of 'structure accepted'.

A similar relationship between structure and architecture is seen in the Public Library

Building at Munster in Germany by Boles and Wilson (Fig. 2.11). The building is in two linked parts which are separated by a pedestrian street. Each part has a relatively straightforward section but is more complex in plan (Fig. 2.12). The juxtaposition of solid and void on the exterior walls is also fairly complicated. Scrutiny of the plans reveals, however, that the structural make-up of the building is straightforward. In each part a reinforced concrete frame has been used. In the case of the part

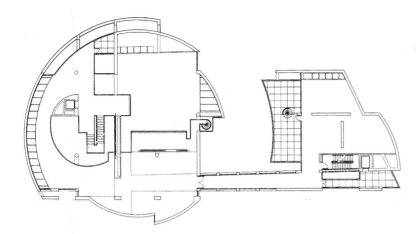

Fig. 2.14 Plan at level three of the Exhibition and Assembly Building, Ulm, Germany, 1993. Richard Meier, architect. The structural plan of this building is based on a 9 m grid. Columns are placed on the grid points and at the locations where the grid lines intersect the perimeter. Beams run between these except where this would carry them across a void. The structural arrangement is simply an adaptation of the basic two-way-spanning skeleton frame layout (Fig 4.51b).

with the curved exterior wall, for example, four lines of vertical structure are used to support the floors, one in each perimeter wall and two running the length of the interior of the building. These lines of vertical structure form a series of rigid frameworks between which the floors span. The structural make-up of the other half of the building is similar. These post-and-beam frameworks are therefore adaptations of the standard one-way-spanning frame plan and constitute a sensible structural solution in view of the spans and loadings involved. The relationship between structure and architecture is one of 'structure accepted'.

The Exhibition and Assembly Building at Ulm in Germany by Richard Meier (Fig. 2.13) is yet another example of a complex form which is based on a reinforced concrete framework. As in the case of the Munster Library the building is in two linked parts. The more complicated of these has a plan which is based on concentric circles. The arrangement is further complicated by the juxtaposition of solid and void which occurs in both the external walls and in the floors of the building.

The structure is based on a 9 m span grid (Fig. 2.14). Columns are placed on the grid points and at the locations where the grid intersects the concentric circles which define the building's perimeter. Beams run between the grid points, except where they would cross voids, and carry the floor slabs. This produces a pattern of support which is sufficient to carry

all areas of the floors. The structural continuity of reinforced concrete is required, however, to accommodate the complex plan-form of the floors. The square column grid would normally be used in conjunction with a flat-slab floor structure. The use of deep beams in this case was due to the lack of structural continuity caused by the large areas of void in the floor structures. The degree of adaptation of the standard form was therefore greater in this building than in the examples described above.

2.6 Conclusion

The buildings described above give an insight into the process by which the design of an architectural structure is carried out. The various aspects of the decision-making process by which the form of structure for a building is determined may be summarised as follows. First, there are several significant global factors which influence the choice of structure type. Perhaps the most important group of these are the architectural objectives being set by the architect(s). The building form which results from the deconstructionist approach followed by, for example, Coop Himmelblau, is likely to be very different from those whose approach is closer to the modernist mainstream, such as Meier or Boles and Wilson. Another form is likely to result from high-tech modernists such as Rogers or Grimshaw. The relationship which exists between the structural design and the architectural design, as discussed in Section 2.2, is yet another global factor which is likely to influence the final form of the structure.

Factors particular to the building being designed will obviously exert a strong influence on its structural make-up. These include the conditions at the site, the scale of the building and the character of its internal layout as determined by the types of accommodation which are required. All of the above factors influence the choice of structural material and the basic structure type. These choices are made from the vocabulary of basic structural forms which is outlined in the following chapters. Once the basic type of structure has been selected a process of adaptation and modification must occur, as described in the examples given above, so that it may satisfy the individual requirements of the particular building.

Chapter 3

Steel structures

3.1 Introduction

Steel is a material which has excellent structural properties. It has high strength in tension and compression and is therefore able to resist bending and axial loads with equal facility; it is the strongest of the commonly used structural materials, being approximately twenty times stronger than timber and ten times stronger than concrete. It is therefore used to make the tallest buildings and the enclosures with the longest spans. Its most common application, however, is for building frames of moderate span, in a wide variety of configurations.

3.2 The architecture of steel – the factors which affect the decision to select steel as a structural material

3.2.1 The aesthetics of steel

The visual vocabulary which is associated with steel structures contains some of the most powerful images of modern architecture. The glass-clad framework (Fig. 3.6), the use of slender, precisely crafted structural components as visual elements (Figs 3.1 and 3.9) and the celebration of structural virtuosity in the form of either breathtakingly long spans (Fig. 3.10) or very tall buildings (Fig. 3.11) are all different aspects of the expressive, and impressive, possibilities of steel. These aesthetic devices have been used from the beginning of the modern period and are still part of the twentieth-century architectural palette. They are often the primary reason for the selection of steel as the structural material for a building.

Steel became available as a building material in the second half of the nineteenth century, following the development of economical processes for its manufacture, and, although its expressive possibilities were not used initially, it was quickly absorbed into the well-established tradition of metal-frame building which had arisen in connection with the use of cast and wrought iron (Fig. 3.2). Iron

Fig. 3.1 Channel 4 Headquarters, London, England, 1994. Richard Rogers & Partners, architects. Concepts such as neatness, precision and high quality control are readily conveyed by the use of an exposed steel structure. [Photo: E. & F. McLachlan]

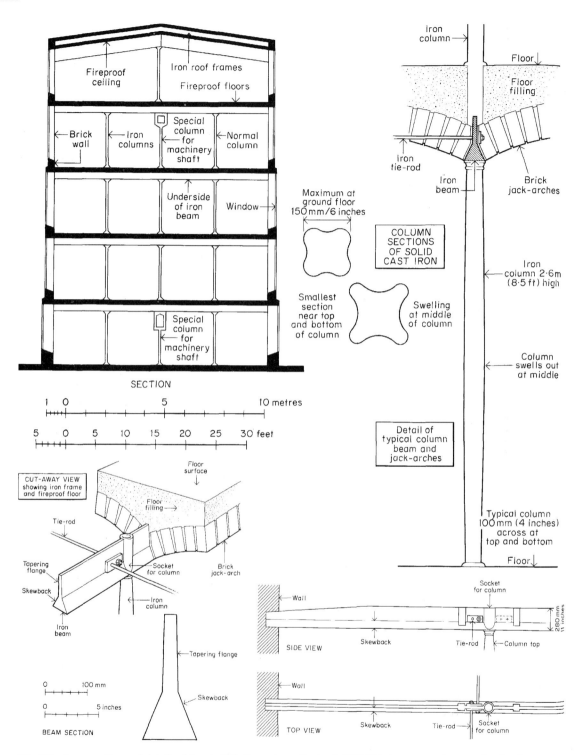

Fig. 3.2 Cross-section and details of Bage's Mill, Shrewsbury, England, 1796. This is a very early example of an iron-frame industrial building. The floors consisted of brick jack-arches, topped by a non-structural filling, supported on a skeleton framework of cast iron beams and columns. No walls were required in the interior. Columns were not provided in the perimeter walls which were therefore of loadbearing masonry. [Illustration, *Mitchell's History of Building*]

frameworks had been developed in the early nineteenth century principally for industrial buildings such as factories, warehouses and railway stations but were little used in the types of building which, at that time, were considered worthy of the descriptive term Architecture. As the century progressed, however, metal frameworks were gradually absorbed into the world of architecture. They were used mainly for 'new' types of building, such as department stores and multi-storey offices, where they allowed uncluttered interiors to be created within buildings which were of conventional external appearance. In Europe, this was the age of revivals and the streets of the commercial districts of late nineteenth-century European cities became filled with imitation Greek and Roman temples and Italian Renaissance palaces which were in fact steel frameworks clad in masonry and which, in their scale and internal arrangements, bore little resemblance to their historic predecessors. In North America, the technology of the steel frame and the existence of a rapidly expanding economy generated a new type of building, that of the skyscraper, which found architectural expression in the work of Louis Sullivan.

The structural technology on which these buildings were based was that of the skeleton framework, a characteristic of which was that loads were channelled into slender elements of low volume in which stress levels were high. Large-scale interior spaces were created with minimal interference from vertical structure, and plan arrangements were varied between levels in multi-storey buildings, within the constraints of a regular column grid, because walls were simply non-loadbearing partitions. The frameworks themselves were normally hidden. Internal columns were cased in fire-proofing materials and finished frequently as one of the classical orders of architecture within a conventional scheme of interior decoration (Fig. 3.3). External walls were of loadbearing masonry and were fashioned into historic stylisations. Eventually, the frame was to take over the structural function of the external walls, although, in the majority of

cases, the walls would continue to be of masonry until well into the twentieth century.

The first steel-framed building in Britain with a non-loadbearing external wall was the Ritz Hotel, of 1903–06, in London's Piccadilly. It is often said that the reason for the adoption of the elaborate stone facade, which did not bear any load, was that the London Building Bye-Laws did not permit otherwise but, given the conservativeness of the architectural establishment in England at the time, it is difficult to imagine that a building of less conventional appearance would have been erected on this very prominent London site.

The constructional system which was developed in the nineteenth century in connection with metal-framed buildings, that of the complete separation of functions between structural framework and non-loadbearing walls, was to become the standard pattern for the steel-frame architecture of the twentieth century. Structural skeletons would be made to support external cladding of many different materials and the building-type would offer architects new opportunities for architectural expression. Masonry, which has ideal properties as an external walling material, would still be used but other, more fragile materials, such as sheet metal, glass and plastics, would also become part of the architectural vocabulary.

The promotion of the steel framework from the status of a purely supportive role to that of a major contributor to the aesthetics of a building was accomplished by the Modernist architects of the early twentieth century. It was inevitable that they would find the material exciting. In the 1920s and 1930s they were engaged in the project of inventing a new architectural vocabulary for the modern world of industry and technology, and, from their point of view, this new structural material had many virtues. Perhaps the greatest was that it was new, but it was also appealing because it was a product of a complex and sophisticated industrial process which made it an appropriate medium with which to develop an architectural vocabulary celebrating industry and technology. Steel also had excellent structural properties, especially high strength in both

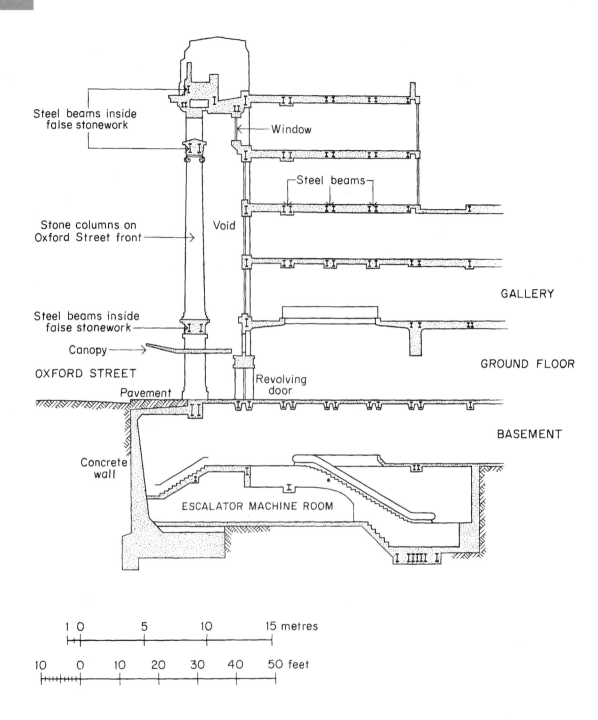

Steel beams inside
false stonework

Window

Steel beams

Stone columns on
Oxford Street front

Void

GALLERY

Steel beams inside
false stonework

Canopy

OXFORD STREET

Pavement

Revolving
door

GROUND FLOOR

BASEMENT

Concrete
wall

ESCALATOR MACHINE ROOM

1 0   5   10   15 metres

10   0   10   20   30   40   50 feet

Fig. 3.3 Part cross-section of Selfridge's Department Store, London, England, 1907. The structure of this building is a skeleton framework of steel beams and columns which support reinforced concrete floor slabs. Unlike in the mill building of Fig 3.2, the exterior walls do not support the floors and consist principally of glazing. The elaborate giant-order columns with their bases and stylised entablature are a non-structural screen designed to give the appearance of a Beaux Arts style 'palace'. [Illustration, *Mitchell's History of Building*]

tension and compression. Its deficiencies – poor durability, poor performance in fire, the difficulty of shaping it into useful components, the high weight of the components – were not significant, and would be overcome by employing other technologies. The large ecological cost, in terms of transportation of raw materials and of the high energy consumption and pollution associated with its manufacture, were not an issue at the time. So far as the modernists were concerned, here was a new and exciting material whose expressive possibilities, if explored, might lead to a truly appropriate architectural vocabulary for the twentieth century.

The technology of the steel framework contributed to the aesthetics of architecture in the 1920s and 1930s in two quite separate ways. Firstly, it was crucial to the development of the glass-clad building; secondly, it made tenable the overt use of structural elements as constituents of a modern visual vocabulary. These two aspects of the aesthetics of Modernism were, and still are, often combined by architects and confused by critics. They are, however, different and distinct aspects of the relationship between architecture and the structural technology of steel frameworks.

The aesthetic programme of the glass-clad framework is concerned with 'transparency'. This, and the use of 'crystalline' form, were given symbolic meaning in the 1920s by the Expressionists, notably Bruno Taut. One of the most striking images to be published by this group, however, was the well-known glass skyscraper project of Mies van der Rohe (Fig. 3.4). Despite its Expressionist genesis this building form survived the gradual triumph of Rationalism – the Rationalists could regard the glass-clad frame as a logical and honest reduction of the elements of a tall building to its bare essentials – and was used at every scale in the architecture of Modernism, from small domestic buildings to the corporate skyscrapers which dominate the skylines of most capital cities.

Among the visual sources of this architectural vocabulary of glass and steel were the iron-framed warehouses, factories and railway

Fig. 3.4 Glass skyscraper project, 1922, model. Mies van der Rohe, architect. In the early twentieth century many visionary architects considered mass-produced high quality glass to be the ultimate modern building material, principally because it allowed the exterior of a building to be 'dematerialised'. The almost featureless high-rise structure faced in glass was an important architectural innovation of the early modern period.

Fig. 3.5 Crystal Palace, London, England, 1851. Joseph Paxton, architect/engineer. This glass-clad framework made a significant contribution to the visual vocabulary of the architecture of the twentieth century. Unlike the later buildings which it inspired, the arrangement was fully justified here on technical grounds.

stations of the nineteenth century. These had largely been ignored in their own time by the architectural mainstream but were destined to exert great influence on the visual vocabulary of twentieth-century Modernism – rather as the motifs of the architecture of Roman antiquity, as interpreted by the scholars of the Italian Renaissance, had laid the foundations for Neoclassicism and the Beaux-Arts school. A well-known example was the famous Boat Store at Sheerness Dockyard in England (1858–66) in which aspects of twentieth-century Modernism were anticipated, but other types of industrial building, especially the large train-sheds of the railway termini, were also important. Extreme versions of this type of building were the glasshouses – glass-clad frameworks of timber and iron – which had been developed for horticultural purposes in the late eighteenth and early nineteenth centuries. This particular tradition of frame building reached its climax with the Crystal Palace in London (Fig. 3.5), which was built to house the Great Exhibition of 1851.[1] It was virtually a cathedral in glass and iron and was

destined to exert a profound influence on the architecture of the twentieth century.

It is appropriate, in a book which is concerned principally with technical matters, to dwell briefly here on the relationship between technology and aesthetics in the context of the glass-clad framework. It was, of course, principally the *look* of the frame buildings of the nineteenth century, and the Crystal Palace in particular, rather than the novelty of their technology of construction, which appealed to the architects of the Modern Movement. The buildings did, in fact, have a number of technical deficiencies. The poor thermal and acoustic properties of the external walls and the poor durability of the junctions between the individual panes of glass in the transparent skins were problematic. The latter caused leaks to develop, which in turn brought about a deterioration in the condition of the structural frameworks of timber and cast iron. None of these was a particularly serious consideration in the context of most of the glass-clad frameworks of the nineteenth century. The Crystal Palace, for example, was a temporary building designed to house a very large exhibition of short duration, and, from a technical point of view, the glass-clad framework was actually an ideal, and probably inevitable, solution to the problem posed,

---

1 The designer of the building was the landscape gardener Joseph Paxton.

especially as there was a requirement that the building be erected very quickly. The glass-clad metal framework, as a genus of building type, also performed well in the context of train-sheds, where generous provision for daylighting and natural ventilation were essential and where high levels of thermal and acoustic insulation were not required. The indifferent weathertightness of the external envelope was tolerable in the context of a large railway station where a programme of maintenance was accepted as normal.

The technical deficiencies of the glass-clad framework were serious drawbacks in the context of other types of buildings, however, and especially of those within which a 'well tempered environment' was a reasonable expectation. The buildings in which people lived, worked, became educated or were ill – the glass-clad houses, offices, schools and hospitals of the modernist twentieth century, in other words – were seriously deficient in their technical performance. Glass, if used as the sole covering for a building which is intended to be occupied by humans, actually performs rather badly in a technical sense. A masonry wall, which is pierced with glass windows for light and ventilation, is a much better technical solution to the problem of cladding a frame building.

The early Modernists, however, despite their declared allegiance to the idea of celebrating technology, were more interested in aesthetics than in technical performance. Both the Rationalists[2] and the Expressionists, were dealing in metaphor and the symbolic attractiveness of the glass-clad framework ensured that mundane, practical considerations were overridden. The glass-walled building became one of the clichés of twentieth-century Modernism – a triumph of ideas and aesthetics

Fig. 3.6 The Seagram Building, New York, USA, 1957. Ludwig Mies van der Rohe, architect. This was the archetypal glass-clad framework of the late modern period.

over that which was practical in building terms (Fig. 3.6). That triumph was to make a significant contribution to the alienation which subsequently developed between the world of architecture (architects and their apologist critics) and the mainstream of society.

It is worth noting that, although the technical efficacy of using glass cladding may be

---

2 In the context of architecture the terms Rational or Rationalism should never be taken literally. They did not mean 'that which is logical, sensible and practical'. The Rationalists produced an architectural vocabulary which symbolised the *idea* of being logical, sensible and practical.

Fig. 3.7 German Pavilion, World Exhibition, Barcelona, 1929. Ludwig Mies van der Rohe, architect. The slender steel structural columns, faced in thin coverings of stainless steel, made a significant contribution to the aesthetics of this seminal building.

questioned, the *structural* performance of the glass-clad steel frameworks of the twentieth century was normally satisfactory. Most of the buildings fall into the category of 'structure accepted'[3] in which the technical performance of the structure was not compromised for visual reasons.

In many of the buildings discussed above, and especially the multi-storey buildings, the steel which formed the loadbearing structure was not actually visible. It was hidden, encased in fire-proofing materials for good technical reasons, and its presence was acknowledged only in the architectural treatment. It is for this reason that a distinction is made here between the role of steel in the glass-clad framework, where it served a purely structural function which happened to make a new aesthetic possible, and the other aesthetic part which it has played in modern architecture, that of a distinctive element in a visual vocabulary.

Metal components can be shaped with great precision and this, together with the low volume of the elements in skeleton frameworks of steel, can be used to create structures of great elegance. The exposure of steelwork and its incorporation into the visual vocabulary is an aesthetic device which has been used extensively in the twentieth century. The architectural intention in this case was more-or-less

---

3 See Section 2.2, and Macdonald, *Structure and Architecture*, Chapter 7.

the same as with the glass-clad framework – the creation of an aesthetic which celebrated the idea of progress, of industrial technology and of the modern lifestyle which these made possible – but the method of expression was more overt. The visual and tactile qualities of steel became important factors in the aesthetic make-up of the building.

Again, Mies van der Rohe is one of the principal exemplars. In the Barcelona Pavilion of 1929 (Fig. 3.7) the steel columns act in conjunction with new treatments of traditional materials, such as marble, to create a new aesthetic. In the Farnsworth House of 1946–50 (Fig. 3.8) the I-shaped structural elements of a glass-clad framework are exposed and bring the visual vocabulary of the engineering workshop into the world of the country retreat.

The exposure of steel frameworks for aesthetic reasons was one of the favoured stylistic devices of the so-called 'high-tech' architecture of the 1970s and 1980s (Fig. 3.9). These buildings, which also were frequently glass-clad frameworks, were visually spectacular and among the most memorable architectural images of the late twentieth century.

The relationship between the structure and the architecture in all of these cases was one of 'structure symbolised'[4] and this way of treating structure usually had the paradoxical effect of compromising its technical performance, because it resulted in structural forms being manipulated predominantly in accordance with visual rather than with technical criteria.

Other problems associated with the exposure of steelwork were those of maintenance of the structure and of fire protection. The second of these arose only in the case of multi-storey structures and was the principal reason why steel structures were rarely exposed other than in single-storey buildings. A notable exception was the Centre Pompidou in Paris, where fairly elaborate fire-protection systems were provided to ensure that the steelwork which was exposed on the exterior of the building would meet the required fire

Fig. 3.8 The Farnsworth House, Illinois, USA, 1951. Ludwig Mies van der Rohe, architect. The mass-produced structural steel beam and column I-sections form important elements in the aesthetic vocabulary of this building.

Fig. 3.9 Sainsbury supermarket, London, England, 1986–89. Nicholas Grimshaw & Partners, architects. Exposed steel structures with exaggerated connections were one of the prominent features of the 'high-tech' style. [Photo: E. & F. McLachlan]

Fig. 3.10 Hong Kong Stadium, Hong Kong, 1995. Ove Arup & Partners, structural engineers. The principal structural elements of this long-span structure are steel arches which have a space-framework configuration. [Photo: Ove Arup Partnership]

performance. All of the structural steelwork in the interior of this building was covered with insulating material and was therefore not exposed and visible.

Steel has made an important contribution to the development of two other categories of building, namely, the very tall building and the very long-span enclosure. Where steelwork has been used in these contexts the limits of what was technically possible were sometimes approached. In such situations the forms which were adopted had to be determined principally from technical considerations. In these two types of steel building, structural requirements dominated the form which was adopted and the resulting building was often spectacular and exciting in appearance. These buildings demonstrate the features of 'true structural high tech'[5] which should not be

confused with the technically misleading version described above, which might be termed 'stylistic' high tech.

In the long-span structure (Fig. 3.10) the level of efficiency (strength-to-weight ratio) must be high. The types of steel structure which have been developed to achieve this are the space framework, often in arched or barrel-vaulted configuration, and the cable network (see Sections 3.6.2.3 and 3.6.4).

In the very tall building (Fig. 3.11) the critical structural problem is that of providing adequate resistance to lateral loading. The most efficient solution to this is to place the structure on the perimeter of the plan, and thus treat the building as a hollow, cantilevered tube. In such circumstances the structure is concentrated in the exterior walls with the result that it inevitably affects the appearance of the building.

In the late twentieth century, the introduction of the ideas of Deconstruction into archi-

5 See Section 2.2.

tecture has led to the extension of the visual vocabulary of glass and steel. In the buildings of architects such as the Coop Himmelblau group (Fig. 2.1) and Behnisch (Fig. 2.2) this most 'technical' of materials has been used in the 'structure ignored' type of relationship to support buildings with highly complex and irregular geometries. It is the very high strength of steel in both tension and compression and its ability to sustain high levels of concentrated load which makes these free-form networks of elements possible.

To sum up, the use of steel as the structural material creates a range of aesthetic opportunities for the designer of a building, any one of which might provide the justification for its selection. Firstly, steel provides a skeleton-frame structure which allows the freedom in internal planning and external treatment which is associated with buildings in which the walls are non-structural. An extreme instance of this is the building with transparent glass walls, but the freedom can be exploited in other ways in the treatment of both the interior and the exterior of the building. Secondly, the great strength of steel allows the creation of very slender structural elements. This, together with walls of light-weight appearance, can be used to create a feeling of lightness, openness and structural elegance. Thirdly, the variety of precisely shaped elements which steel makes possible and the 'technical' ambience which surrounds these allows them to be used to convey symbolic meaning. This possibility has so far been used in a straightforward manner to express the idea of technical progress. It could, of course, be used ironically to convey a quite different meaning and there is perhaps evidence that this is happening in the architecture of Deconstruction.

Thus, the aesthetic of steel is the aesthetic of the structural framework in which the elements are slender and in which the problem of providing support is solved in an apparently effortless manner. It has normally also been the aesthetic of the celebration of technology, the machine and the production line as they are symbolised by straight-edged forms, regular and often rectilinear grids and overtly machine-like detailing.

Fig. 3.11 Sears Tower, Chicago, USA, 1974. Skidmore, Owings & Merrill, architects. The Sears tower, which was for over two decades the tallest building in the world, has a steel framework structure. [Photo: I. Boyd Whyte].

3.2.2 The technical performance of steel as a structural material

3.2.2.1 Introduction
Steel is the strongest of the four commonly used structural materials but has a strength-to-weight ratio which is similar to that of timber (i.e. very

high). It is ideally suited to the form of skeleton frameworks, where the principal alternative in multi-storey buildings is reinforced concrete. Its high ratio of strength to weight also makes it suitable for lightweight frameworks such as are used in roof structures. In this application the principal alternative is timber. The advantages and disadvantages of steel in relation to these materials is reviewed briefly here.

3.2.2.2 Advantages of steel

Strength
The high strength of steel, and its high ratio of strength to weight, makes it suitable for use in single- and multi-storey skeleton frames, over a large range of spans and building heights. The material therefore offers the advantages of skeleton-frame construction in all of its manifestations. In addition to freedom from the restrictions of loadbearing walls in the planning of both the interior and the exterior of the building, as mentioned above, these include the subsequent flexibility to alter plans and elevations when required.

Considerable flexibility is also possible in the planning of the building services and in their subsequent maintenance and replacement when this becomes necessary. These advantages are, of course, also present with reinforced concrete frameworks but because the structural elements in steel frames are more slender and less obtrusive than those in equivalent reinforced concrete frames, the use of steel allows interiors with lighter and more open aspects to be created.

Ratio of strength to weight
Steel frames are lighter than reinforced concrete frames of equivalent strength, particularly if efficient types of element such as triangulated girders are used. This makes them more suitable than reinforced concrete frames for use in single-storey buildings and the roof structures of multi-storey buildings. In this role the principal alternative to steel is timber.

Quality control
Steel is manufactured under conditions of strict quality control and its properties can be relied upon to be within narrow specified limits; this allows relatively small factors of safety to be adopted in the structural design calculations and is a further reason why light and slender elements are possible.

Appearance
Due to the strict quality control which is exercised during its manufacture and to the methods which are used in the final shaping of steel components, the finished structure has a distinctive appearance which is characterised by slender elements, smooth surfaces and straight, sharp edges.

Prefabrication
Steel structures are assembled from prefabricated components which are produced off-site and this allows their dimensions and general quality to be carefully controlled. It also results in fast erection of the structure *on* site and enables a relatively simple erection process to be adopted, even on difficult sites. Prefabrication with site-jointing also means that it is relatively easy for the designer to exercise control over whether or not the structure is statically determinate.

3.2.2.3 Disadvantages of steel

Intractability
Steel is a very tough material which is difficult to work and shape in the solid form and this has a number of consequences. It means that, in most steelwork design, it is necessary to specify elements from a standard range of components which are produced by steel manufacturers and to carry out the minimum amount of modification to these. The standard elements, which are produced by a hot-rolling or cold-forming process (see Section 3.4), are straight and parallel sided with the result that steel frameworks must normally have a regular, rectilinear, or at least straight-edged geometry.

The production of 'tailor-made' cross-sections, or of geometries which are curvilinear, is difficult and the use of steelwork tends therefore to place more restrictions on the overall forms of structures than does the use of reinforced concrete. Also, the final adjustment of the dimensions of elements on site is difficult, if not impossible, so that a much higher standard of quality control is required in such processes as the initial shaping and cutting to length, than is necessary with, for example, timber.

Weight
The density of steel is high and this makes individual components fairly heavy. Elements such as beams and columns are difficult to move around on site and cranes are normally required for the assembly of steel structures.

Cost
The basic cost of a steel structure is normally greater than that of its timber or reinforced concrete equivalent. The shorter on-site construction time can be a compensatory factor, however.

Durability
Most steels are relatively unstable chemically and a corrosion-protection scheme is normally required for a steel structure.

Performance in fire
Steel loses its ability to carry load at a relatively low temperature (around 500 °C) and this means that, while a steel structure does not actually burn, it will collapse in fire unless it is kept cool. This is normally achieved by protecting the steelwork with a suitably thick layer of fire resistant insulating material but sometimes more sophisticated methods, such as water cooling systems, are used. The traditional fireproofing material was concrete – the elements of a steel frame were simply encased in concrete – but much lighter materials which are easier to apply have been developed. The need to provide fireproofing for steelwork nevertheless increases the complexity of a steel-frame building and adds to the cost of the structure.

The need to provide fire protection is obviously particularly problematic if there is an intention to expose the structure as part of its architectural expression. A prominent example already mentioned is the Centre Pompidou in Paris where an elaborate system of shutters, which would operate automatically in the event of a fire, had to be installed adjacent to the glass external wall in order to protect the steelwork on the exterior of the building should a fire occur. The steel trusses in the interior were encased in fireproofing material which was wrapped in a thin skin of sheet metal to preserve the appearance of a steel structure. A similar architectural language was proposed for the Lloyd's Headquarters Building in London. In this case, however, no satisfactory way of meeting the fire-resistance requirements could be found and reinforced concrete was finally adopted as the structural material.

3.2.2.4 Conclusion
Steel is a material whose properties make it particularly suitable for skeleton-frame structures. The principal advantage which results from its adoption is that it releases the designer from the restrictions on internal planning and the aesthetic treatment of the exterior which are imposed by the use of loadbearing walls. In addition, the high strength of steel makes possible a very wide range of frame types; both single-storey and multi-storey frames can be constructed over a very wide range of spans and very tall multi-storey structures are also possible. In all cases the structures will be lighter than equivalent reinforced concrete frames. Other advantages stem from the fact that steel structures are prefabricated: these include speed of erection, ease of assembly on difficult sites and, if required, statical determinacy. Restrictions on the overall form of a building must normally be accepted, however, due to the limited range of components which are readily available, and the detailing of the structure is likely to be complicated by the need to provide fire protection. In addition, the cost of the steel-frame structure is likely to be higher than that of an equivalent reinforced concrete frame.

3.3 The properties and composition of steel

Steel is a ferrous metal (its principal constituent being iron), but it contains a number of other chemical elements which act as alloying agents and which have a critical effect on its properties. The most important of these is carbon; steel is defined as ferrous metal with a carbon content in the range of 0.02% to 2%. Low-carbon steels are relatively soft and ductile while those with a high carbon content are hard and brittle. The range of properties is fairly wide, depending on the precise levels of carbon and of other trace elements which are present. The term 'steel' refers therefore not to a single metal but to a range of alloys.

Structural steels are 'mild steels', which have a carbon content of around 0.23%; the other principal alloying agent is manganese which is maintained at around 1.6%; sulphur and phosphorus are also present. A range of structural steels is available in most countries. Those which are used in the UK are specified in BS 4360 'Weldable Structural Steels' which is currently being superseded by a European Standard EN 10 025. The latter incorporates the provisions of BS 4360.

BS 4360 specifies four grades of structural steel: 40, 43, 50 and 55. The grade numbers refer to the tensile strength values (400, 430, 500 and 550 newtons per square millimetre).[6] Within each grade there are various sub-grades: A, B, C, etc. determined by minor variations in chemical composition, principally carbon content. The higher sub-grades (lower carbon content) have slightly improved mechanical properties and perform better in respect of welding.

The properties of steel can be manipulated by heat treatment. If the metal is cooled very rapidly (quenched) the crystalline structure is quite different from that which results from gradual cooling. Quenched steel is extremely hard and brittle and is not used in structural engineering. Following re-heating, however, the metal regains its ductility and the level of

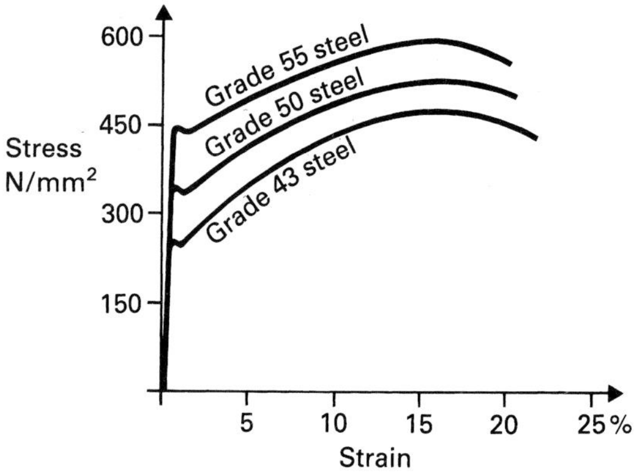

Fig. 3.12 The relationship between stress and strain in typical structural steels. The short section of the graph between the elastic and plastic ranges, in which the graph is more-or-less horizontal, is of fundamental importance in determining the excellent structural behaviour of steel.

brittleness/ductility (and therefore yield strength) which is achieved can be accurately controlled in this process (which is called tempering). Heat treated steels are used in structural engineering only for very specialised applications, the most common of which is in the manufacture of high strength friction-grip bolts.[7]

Steel is a high-strength material which has equal strength in tension and compression; the ultimate strength and design strength values which are used in the UK are given in Table 3.1, which is reproduced from BS 5950 'The Structural Use of Steelwork in Building'.

The relationship between stress and strain of a typical structural steel is shown in Fig. 3.12 and it will be seen that 'elastic' behaviour (linear behaviour in which the graph of stress against strain is a straight line) occurs in the lower part of the load range. In the higher load range the relationship is curved (inelastic, non-linear behaviour) and a larger increase in strain results for a given increase in stress. The location in the graph at which the transition to

6 It is intended to replace these designations with grades based on yield stress values. Thus grade 43A will become 275A as the yield stress for this type of steel is 275 N/mm².

7 Blanc, McEvoy and Plank, *Architecture and Construction in Steel*, Chapter 3.

Table 3.1 Basic design strengths for steel

BS 4360 Grade	Thickness, less than or equal to (mm)	Strength of sections, plates and hollow sections (N/mm²)
43	16	275
A, B and C	40	265
	100	245
50	16	355
B and C	63	340
	100	325
55	16	450
C	25	430
	40	415

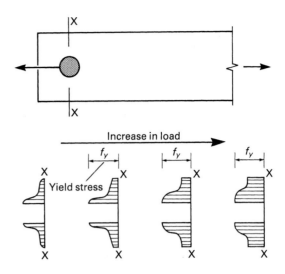

Fig. 3.13 The stress-relieving mechanism. Each of the four diagrams in the lower half of this illustration shows the distribution of stress across the cross-section X–X. In the first diagram all of the material is stressed within the elastic range and the material which is closest to the bolt hole is the most highly stressed. As the load rises the level of stress also rises. Once the yield stress is passed at the most highly stressed locations the horizontal portion of the stress–strain graph is entered and it is possible for the stress level to remain constant while stress levels in other parts of the cross-section continue to rise. The distribution of stress then becomes more even. This is an example of the stress-relieving mechanism which is responsible for the ability of steel to resist high levels of tensile load.

non-linear behaviour occurs is termed the 'yield point'.

The portion of the stress–strain graph which immediately follows the yield point is more-or-less horizontal. This feature illustrates a very important property of steel which is the mechanism for the relief of stress concentrations, especially in the vicinity of connections. If, for example, one part of a cross-section tends to be highly stressed relative to other parts (this might occur in the vicinity of a bolt), the highly stressed material could reach its yield point while the average stress was still relatively small (Fig. 3.13). If more load were applied the strain would increase equally in all parts of the cross-section but the level of stress in the most highly stressed area would tend to remain constant because the horizontal portion of the stress–strain graph would have been entered by the material in that area. In other parts of the cross-section, where the stress was within the elastic range, the stress level would continue to increase with increasing strain, however, and the distribution of stress would therefore tend to become more even. The existence of the short horizontal portion of the stress/strain graph is therefore a stress-relieving feature. It is a very important factor in determining the good structural performance of steel.

Another significant aspect of the stress–strain graph is the amount of deformation which occurs before failure. A very large amount of deformation is in fact required before steel fractures. This too is a safety feature because it means that an overloaded structure will suffer a large deflection which gives warning of impending collapse.

3.4 Structural steel products

3.4.1 Introduction
Three basic techniques are used to form metal components. These are casting, in which molten metal is poured into a suitably shaped mould, forging, where solid metal is beaten or otherwise forced into a particular shape, and machining, where material is cut away by various means from a basic block of metal, to form a component with a particular shape. All

Fig. 3.14 Renault Sales Headquarters, Swindon, England, 1983. Foster Associates, architects; Ove Arup & Partners, structural engineers. Steel castings are used for the jointing components of complex shape which occur at the ends of the tie bars in this connection. [Photo: Alastair Hunter]

three methods are used in the formation of steelwork elements. The greatest tonnage of elements for structural steelwork is produced by forging; two processes are used, hot-rolling and cold-forming[8] and large ranges of steel components are made by both of these processes. The casting of major steel components is rare and the principal use of this process is as the preliminary stage in the forming of small jointing components (Fig. 3.14). Machining is the most precise of the shaping processes and is used both in the final stages of production of small components, such as nuts and bolts, and in the final preparation of large steel elements for a particular

structure. The drilling of bolt holes, for example, is a machining process.

3.4.2 Hot-rolled sections

Hot-rolling is a forging process in which hot billets of steel are passed repeatedly between profiled rollers to produce straight elements which have particular shapes and sizes of cross-section (Fig. 3.15). Several sets of rollers are normally required to transform a rough billet into a finished element with a cross-section which has satisfactory structural properties. Most sets of rollers are capable of

---

8 See Blanc, McEvoy and Plank, *op.cit.*, Chapters 4 and 5 for a description of the manufacturing processes of steel components.

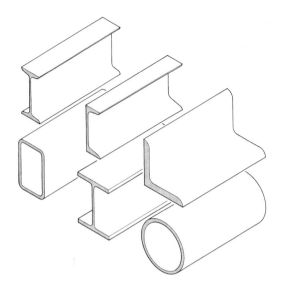

Fig. 3.15 Hot-rolled steel sections. These mass-produced elements are formed by a rolling process which results in the characteristic parallel-sided arrangement with constant cross-section. The thickness of the metal is relatively high and results in correspondingly high load-carrying capacity.

producing only one cross-section; some can be adjusted within narrow limits to produce a small range of cross-section weights within a particular overall cross-section size (serial size).

Steel manufacturers produce a limited range of cross-sectional shapes by this process; the principal ones are as follows:

I-sections

These are typified by the UK universal beam series (Table 3.2); they have cross-sectional shapes with a high second moment of area[9] about one principal axis and a much lower second moment of area about the other principal axis. They are intended for use as beam elements, which are normally subjected to bending in one plane only. The finishing rollers which are used to produce universal beams

consist of a complex arrangement of two pairs of rollers, one with horizontal and one with vertical axes. This allows the thickness of the flanges and webs to be varied within small limits and accounts for the different weights of section which are produced within each serial size.

H-sections

These are typified by the UK universal column range (Table 3.3) and have more-or-less the same second moment of area about both principal axes. They are designed to resist axial compression, which requires a cross-section with more-or-less equal rigidity about all axes to prevent buckling.

Channel and angle sections

These are much smaller than the I and H ranges and are used mainly for secondary elements, such as purlins (secondary elements in roof structures), and for the sub-elements of steel trusses and built-up cross-sections.

Hollow sections

These ranges are produced in circular, square and rectangular form. The lighter sections are used mainly for trusses and space frames. Heavier sections are used as beams and columns.

Flat plate

This is available in thicknesses from 5 mm to 65 mm and is used for elements with built-up cross-sections (plate girders and compound beams) and for jointing components.

Strip

This is produced in the form of thin flat sheet. Its principal use in structural engineering is as the raw material for cold-formed sections.

Wire, rope and cable

Steel wire is manufactured by drawing solid bar through a series of small diameter dies accompanied by heat treatment. Wires (usually around 50) are spun into strands. A rope consists of a number of strands (usually 6) spun around a core of steel or fibre. A cable is made up of six ropes. The largest ropes are 100 mm in diameter and cables may be up to 250 mm in diameter.

9 See Macdonald, *Structure and Architecture*, Appendix 2, for an explanation of this term.

Table 3.2 Dimensional properties of universal beam sections

To: BS4 Part 1

Designation Serial size mm	Mass per metre kg	Depth of section D mm	Width of section B mm	Thickness Web t mm	Thickness Flange T mm	Area of section cm²	Moment of inertia Axis x–x cm⁴	Moment of inertia Axis y–y cm⁴	Radius of gyration Axis x–x cm	Radius of gyration Axis y–y cm	Elastic modulus Axis x–x cm³	Elastic modulus Axis y–y cm³	Plastic modulus Axis x–x cm³	Plastic modulus Axis y–y cm³
914 × 419	388	920.5	420.5	21.5	36.6	494.5	718742	45407	38.1	9.58	15616	2160	17657	3339
	343	911.4	418.5	19.4	32.0	437.5	625282	39150	37.8	9.46	13722	1871	15474	2890
914 × 305	289	926.6	307.8	19.6	32.0	368.8	504594	15610	37.0	6.51	10891	1014	12583	1603
	253	918.5	305.5	17.3	27.9	322.8	436610	13318	36.8	6.42	9507	871.9	10947	1372
	224	910.3	304.1	15.9	23.9	285.3	375924	11223	36.3	6.27	8259	738.1	9522	1162
	201	903.0	303.4	15.2	20.2	256.4	325529	9427	35.6	6.06	7210	621.4	8362	982.5
838 × 292	226	850.9	293.8	16.1	26.8	288.7	339747	11353	34.3	6.27	7986	772.9	9157	1211
	194	840.7	292.4	14.7	21.7	247.2	279450	9069	33.6	6.06	6648	620.4	7648	974.4
	176	834.9	291.6	14.0	18.8	224.1	246029	7792	33.1	5.90	5894	534.4	6809	841.5
762 × 267	197	769.6	268.0	15.6	25.4	250.8	239894	8174	30.9	5.71	6234	610.0	7167	958.7
	173	762.0	266.7	14.3	21.6	220.5	205177	6846	30.5	5.57	5385	513.4	6197	807.3
	147	753.9	265.3	12.9	17.5	188.1	168966	5468	30.0	5.39	4483	412.3	5174	649.0
686 × 254	170	692.9	255.8	14.5	23.7	216.6	170147	6621	28.0	5.53	4911	517.7	5624	810.3
	152	687.6	254.5	13.2	21.0	193.8	150319	5782	27.8	5.46	4372	454.5	4997	710.0
	140	683.5	253.7	12.4	19.0	178.6	136276	5179	27.6	5.38	3988	408.2	4560	637.8
	125	677.9	253.0	11.7	16.2	159.6	118003	4379	27.2	5.24	3481	346.1	3996	542.0
610 × 305	238	633.0	311.5	18.6	31.4	303.8	207571	15838	26.1	7.22	6559	1017	7456	1574
	179	617.5	307.0	14.1	23.6	227.9	151631	11412	25.8	7.08	4911	743.3	5521	1144
	149	609.6	304.8	11.9	19.7	190.1	124660	9300	25.6	6.99	4090	610.3	4572	936.8
610 × 229	140	617.0	230.1	13.1	22.1	178.4	111844	4512	25.0	5.03	3626	392.1	4146	612.5
	125	611.9	229.0	11.9	19.6	159.6	98579	3933	24.9	4.96	3222	343.5	3677	535.7
	113	607.3	228.2	11.2	17.3	144.5	87431	3439	24.6	4.88	2879	301.4	3288	470.2
	101	602.2	227.6	10.6	14.8	129.2	75720	2912	24.2	4.75	2515	255.9	2882	400.0
533 × 210	122	544.6	211.9	12.8	21.3	155.8	76207	3393	22.1	4.67	2799	320.2	3203	500.6
	109	539.5	210.7	11.6	18.8	138.6	66739	2937	21.9	4.60	2474	278.8	2824	435.1
	101	536.7	210.1	10.9	17.4	129.3	61659	2694	21.8	4.56	2298	256.5	2620	400.0
	92	533.1	209.3	10.2	15.6	117.8	55353	2392	21.7	4.51	2076	228.6	2366	356.2
	82	528.3	208.7	9.6	13.2	104.4	47491	2005	21.3	4.38	1798	192.2	2056	300.1

Designation	Mass (kg/m)	D	B	t	T	Area	I_{xx}	I_{yy}	r_x	r_y	Z_x	Z_y	S_x	S_y
457 × 191	98	467.4	192.8	11.4	19.6	125.3	45717	2343	19.1	4.33	1956	243.0	2232	378.3
	89	463.6	192.0	10.6	17.7	113.9	41021	2086	19.0	4.28	1770	217.4	2014	337.9
	82	460.2	191.3	9.9	16.0	104.5	37103	1871	18.8	4.23	1612	195.6	1833	304.0
	74	457.2	190.5	9.1	14.5	95.0	33388	1671	18.7	4.19	1461	175.5	1657	272.2
	67	453.6	189.9	8.5	12.7	85.4	29401	1452	18.5	4.12	1296	152.9	1471	237.3
457 × 152	82	465.1	153.5	10.7	18.9	104.5	36215	1143	18.6	3.31	1557	149.0	1800	235.4
	74	461.3	152.7	9.9	17.0	95.0	32435	1012	18.5	3.26	1406	132.5	1622	209.1
	67	457.2	151.9	9.1	15.0	85.4	28577	878	18.3	3.21	1250	115.5	1441	182.2
	60	454.7	152.9	8.0	13.3	75.9	25464	794	18.3	3.23	1120	103.9	1284	162.9
	52	449.8	152.4	7.6	10.9	66.5	21345	645	17.9	3.11	949.0	84.6	1094	133.2
406 × 178	74	412.8	179.7	9.7	16.0	95.0	27329	1545	17.0	4.03	1324	172.0	1504	266.9
	67	409.4	178.8	8.8	14.3	85.5	24329	1365	16.9	4.00	1188	152.7	1346	236.5
	60	406.4	177.8	7.8	12.8	76.0	21508	1199	16.8	3.97	1058	134.8	1194	208.3
	54	402.6	177.6	7.6	10.9	68.4	18626	1017	16.5	3.85	925.3	114.5	1048	177.5
406 × 140	46	403.2	142.4	6.9	11.2	59.0	15647	539	16.3	3.02	777.8	75.7	888.4	118.3
	39	397.3	141.8	6.3	8.6	49.4	12452	411	15.9	2.89	626.9	58.0	720.8	91.08
356 × 171	67	364.0	173.2	9.1	15.7	85.4	19522	1362	15.1	3.99	1073	157.3	1212	243.0
	57	358.6	172.1	8.0	13.0	72.2	16077	1109	14.9	3.92	896.5	128.9	1009	198.8
	51	355.6	171.5	7.3	11.5	64.6	14156	968	14.8	3.87	796.2	112.9	894.9	174.1
	45	352.0	171.0	6.9	9.7	57.0	12091	812	14.6	3.78	686.9	95.0	773.7	146.7
356 × 127	39	352.8	126.0	6.5	10.7	49.4	10087	357	14.3	2.69	571.8	56.6	653.6	88.68
	33	348.5	125.4	5.9	8.5	41.8	8200	280	14.0	2.59	470.6	44.7	539.8	70.24
305 × 165	54	310.9	166.8	7.7	13.7	68.4	11710	1061	13.1	3.94	753.3	127.3	844.8	195.3
	46	307.1	165.7	6.7	11.8	58.9	9948	897	13.0	3.90	647.9	108.3	722.7	165.8
	40	303.8	165.1	6.1	10.2	51.5	8523	763	12.9	3.85	561.2	92.4	624.5	141.5
305 × 127	48	310.4	125.2	8.9	14.0	60.8	9504	460	12.5	2.75	612.4	73.5	706.1	115.7
	42	306.6	124.3	8.0	12.1	53.2	8143	388	12.4	2.70	531.2	62.5	610.5	98.24
	37	303.8	123.5	7.2	10.7	47.5	7162	337	12.3	2.67	471.5	54.6	540.5	85.66
305 × 102	33	312.7	102.4	6.6	10.8	41.8	6487	193	12.5	2.15	415.0	37.8	479.9	59.85
	28	308.9	101.9	6.1	8.9	36.3	5421	157	12.2	2.08	351.0	30.8	407.2	48.92
	25	304.8	101.6	5.8	6.8	31.4	4387	120	11.8	1.96	287.9	23.6	337.8	37.98
254 × 146	43	259.6	147.3	7.3	12.7	55.1	6558	677	10.9	3.51	505.3	92.0	568.2	141.2
	37	256.0	146.4	6.4	10.9	47.5	5556	571	10.8	3.47	434.0	78.1	485.3	119.6
	31	251.5	146.1	6.1	8.6	40.0	4439	449	10.5	3.35	353.1	61.5	395.6	94.52
254 × 102	28	260.4	102.1	6.4	10.0	36.2	4008	178	10.5	2.22	307.9	34.9	353.4	54.84
	25	257.0	101.9	6.1	8.4	32.2	3408	148	10.3	2.14	265.2	29.0	305.6	45.82
	22	254.0	101.6	5.8	6.8	28.4	2867	120	10.0	2.05	225.7	23.6	261.9	37.55
203 × 133	30	206.8	133.8	6.3	9.6	38.0	2887	384	8.72	3.18	279.3	57.4	13.3	88.05
	25	203.2	133.4	5.8	7.8	32.3	2356	310	8.54	3.10	231.9	46.4	259.8	71.39

Table 3.3 Dimensional properties of universal column sections

To: BS4 Part 1

Designation Serial size mm	Mass per metre kg	Depth of section D mm	Width of section B mm	Thickness Web t mm	Thickness Flange T mm	Area of section cm²	Moment of inertia Axis x-x cm⁴	Moment of inertia Axis y-y cm⁴	Radius of gyration Axis x-x cm	Radius of gyration Axis y-y cm	Elastic modulus Axis x-x cm³	Elastic modulus Axis y-y cm³	Plastic modulus Axis x-x cm³	Plastic modulus Axis y-y cm³
356 × 406	634	474.7	424.1	47.6	77.0	808.1	275140	98211	18.5	11.0	11592	4632	14247	7114
	551	455.7	418.5	42.0	67.5	701.8	227023	82665	18.0	10.9	9964	3951	12078	6058
	467	436.6	412.4	35.9	58.0	595.5	183118	67905	17.5	10.7	8388	3293	10009	503.8
	393	419.1	407.0	30.6	49.2	500.9	146765	55410	17.1	10.5	7004	2723	8229	4157
	340	406.4	403.0	26.5	42.9	432.7	122474	46816	16.8	10.4	6027	2324	6994	3541
	287	393.7	399.0	22.6	36.5	366.0	99994	38714	16.5	10.3	5080	1940	5818	2952
	235	381.0	395.0	18.5	30.2	299.8	79110	31008	16.2	10.2	4153	1570	4689	2384
Column core 356 × 368	477	427.0	424.4	48.0	53.2	607.2	172391	68056	16.8	10.6	8075	3207	9700	4979
	202	374.7	374.4	16.8	27.0	257.7	66307	23632	16.0	9.57	3540	1262	3977	1917
	177	368.3	372.1	14.5	23.8	225.7	57153	20470	15.9	9.52	3104	1100	3457	1668
	153	362.0	370.2	12.6	20.7	195.2	48525	17469	15.8	9.46	2681	943.8	2964	1430
	129	355.6	368.3	10.7	17.5	164.9	40246	14555	15.6	9.39	2264	790.4	2482	1196
305 × 305	283	365.3	321.8	26.9	44.1	360.4	78777	24545	14.8	8.25	4314	1525	5101	2337
	240	352.6	317.9	23.0	37.7	305.6	64177	20239	14.5	8.14	3641	1273	4245	1947
	198	339.9	314.1	19.2	31.4	252.3	50832	16230	14.2	8.02	2991	1034	3436	1576
	158	327.2	310.6	15.7	25.0	201.2	38740	12524	13.9	7.89	2368	806.3	2680	1228
	137	320.5	308.7	13.8	21.7	174.6	32838	10672	13.7	7.82	2049	691.4	2298	1052
	118	314.5	306.8	11.9	18.7	149.8	27601	9006	13.6	7.75	1755	587.0	1953	891.7
	97	307.8	304.8	9.9	15.4	123.3	22202	7268	13.4	7.68	1442	476.9	1589	723.5
254 × 254	167	289.1	264.5	19.2	31.7	212.4	29914	9796	11.9	6.79	2070	740.6	2417	1132
	132	276.4	261.0	15.6	25.3	167.7	22575	7519	11.6	6.68	1634	576.2	1875	878.6
	107	266.7	258.3	13.0	20.5	136.6	17510	5901	11.3	6.57	1313	456.9	1485	695.5
	89	260.4	255.9	10.5	17.3	114.0	14307	4849	11.2	6.52	1099	378.9	1228	575.4
	73	254.0	254.0	8.6	14.2	92.9	11360	3873	11.1	6.46	894.5	305.0	988.6	462.4
203 × 203	86	222.3	208.8	13.0	20.5	110.1	9462	3119	9.27	5.32	851.5	298.7	978.8	455.9
	71	215.9	206.2	10.3	17.3	91.1	7647	2536	9.16	5.28	708.4	246.0	802.4	374.2
	60	209.6	205.2	9.3	14.2	75.8	6088	2041	8.96	5.19	581.1	199.0	652.0	302.8
	52	206.2	203.9	8.0	12.5	66.4	5263	1770	8.90	5.16	510.4	173.6	568.1	263.7
	46	203.2	203.2	7.3	11.0	58.8	4564	1539	8.81	5.11	449.2	151.5	497.4	230.0
152 × 152	37	161.8	154.4	8.1	11.5	47.4	2218	709	6.84	3.87	274.2	91.78	310.1	140.1
	30	157.5	152.9	6.6	9.4	38.2	1742	558	6.75	3.82	221.2	73.06	247.1	111.2
	23	152.4	152.4	6.1	6.8	29.8	1263	403	6.51	3.68	165.7	52.95	184.3	80.87

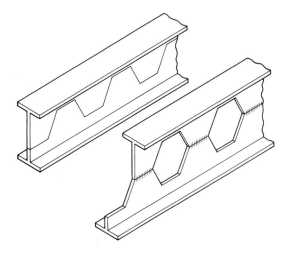

Fig. 3.16 The castellated beam is a modified universal I-section beam. It is more efficient in resisting bending than the parent beam because a higher bending moment can be resisted by the same area of cross-section and therefore volume of material.

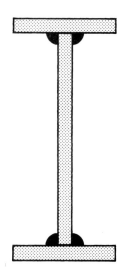

Fig. 3.18 I-section girder assembled by welding from flat plate. These are used when the required size of cross-section is larger than the largest available in the standard range.

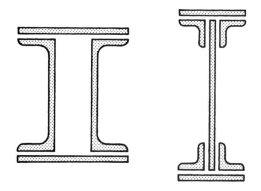

Fig. 3.17 Composite beams, built up by bolting or welding standard sections together, are used where no suitable standard section is available.

The structural sections described above are the basic components from which steel structures are assembled. Elements for particular applications are normally selected from the standard ranges; it is not considered economic to produce 'tailor made' cross-sections by the hot-rolling process due to the high capital cost of the special rolling-mill equipment which would be required. If none of the standard sections is suitable for a particular application non-standard sections can be built up from the standard sections. The 'castellating' process, by which universal beam sections are modified (Fig. 3.16), is one such process and provides a range of beam sections which are lighter than the standard hot-rolled sections (Tables 3.2 and 3.3). Light beam sections can also be manufactured from combinations of channel and angle sections (Fig. 3.17). Where very heavy cross-sections are required, these can be built up from flat plate as welded plate girders (Fig. 3.18).

3.4.3 Cold-formed sections
Cold-formed sections are fabricated from thin sheet steel (strip) by rolling or by folding (press braking), both of which are types of forging. As with hot-rolled sections, manufacturers produce standard ranges, such as angles, channels and I-sections, which are intended to be generally useful for a variety of structural purposes. The capital cost of the equipment required to produce these sections is less than that for hot-rolled products and the ranges which are readily available are much wider. The ease with which steel sheet can be bent into complicated shapes allows

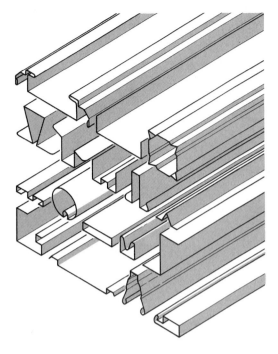

Fig. 3.19 Cold-formed sections. Structural sections can be formed by bending thin sheet. This allows more complex shapes of cross-section to be created than is possible with the hot-rolling process. The metal must be relatively thin, however, so these are lighter sections with a lower carrying capacity than hot-rolled equivalents.

the fabrication of sections with complex geometries (Fig. 3.19) and 'tailor-made' cold-formed sections can be produced for special purposes at a reasonable cost. The sections which are most commonly employed in building are Z-purlins, which are frequently used in conjunction with frames of hot-rolled elements, and profiled-sheet decking. Many kinds of the latter are produced and they are employed both as roof decks and, in conjunction with *in situ* reinforced concrete, as floor structures. Another common use of cold-formed sections is for the flanges of light lattice beams and a number of proprietary ranges of these are available. An example is the Metsec range (Figs 3.20). These are suitable for use as joists in conjunction with loadbearing masonry or with main frames of hot-rolled sections. They are also used to form complete frames in situations where only light loads are carried.

Cold-formed sections are much lighter than hot-rolled sections and they have considerably less strength than the latter. Their principal application has been in single-storey structures; most cold-formed sections do not have

Fig. 3.20 Metsec beam. This is an example of a proprietary beam in which cold-formed sections of relatively complicated shape have been used for the flanges. Elements like this are highly efficient in resisting load and are therefore light. Their limited load-carrying capacity makes them most suitable for use in roof structures.

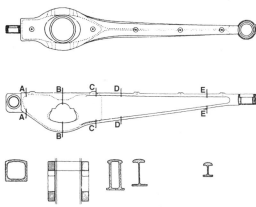

Fig. 3.21 Cast steel 'Gerberette', Centre Pompidou, Paris, France, 1977. Piano and Rogers, architects; Ove Arup & Partners, structural engineers. Casting was the only feasible method by which these large steel components could be manufactured, given their complex shapes in profile and cross-section. Recent developments in casting technique have resulted in an increase in the use of this type of component. [Photo: A. Macdonald]

sufficient strength to be used effectively in multi-storey structures where floor loads must be carried.

3.4.4 Cast steel components

As with other metals, individual components of complex geometry can be produced in steel by casting. In structural engineering this technique is normally confined to the production of small components for joints in structures. Exceptionally, it is used to produce large

structural elements of complex geometry; an example is shown in Fig. 3.21.

3.4.5 Bolts

Two types of bolt are used to make connections in structural steelwork, ordinary bolts and friction-grip bolts. In the case of ordinary bolts the load is transmitted between components through the shanks of the bolts themselves (Fig. 3.22), which are normally loaded in shear but which can be made to

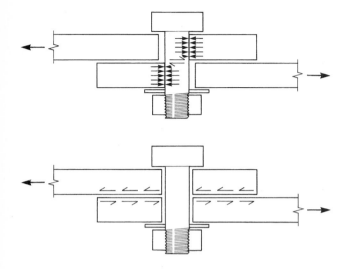

Fig. 3.22 The actions of the two principal types of bolt used in structural engineering are illustrated here.
(a) The 'ordinary' bolt transmits load by bearing and shear. This causes stress concentrations and is not particularly efficient. Connections are simple and cheap to construct by this method, however.
(b) The 'high strength friction grip bolt' clamps together the elements being joined with sufficient pressure to allow the load to be transmitted by friction between the surfaces. This reduces stress concentrations and is a more efficient way of transmitting load through the connection. It is more expensive to construct, however, because both the bolt and the elements being connected must be manufactured to closer tolerances.

carry tension. Ordinary bolts are made in several grades of steel and the two grades which are most commonly used in the UK are 4.6 and 8.8. The first figure in the grade number represents the tensile strength of the steel in $kgf/mm^2 \times 10^{-1}$; the second figure is a factor by which the first is multiplied to give the yield stress of the steel.

Friction-grip bolts operate by clamping components together with such force that the load is transferred by friction on the interface between them. Shear load therefore passes directly from one component to the other and is not routed through the bolts (Fig. 3.22). Friction-grip bolts are also used to carry axial tension through joints, however, in which case the load passes along the bolt shank.

3.5 Performance of steel in fire

The performance of a building in a fire is an important design consideration, the principal concern being the safety of people who are in or near the building at the time. There are many aspects to this including the provision of adequate means of escape and the control of smoke. So far as the structure is concerned, the principal concern is the prevention of instability due to the lowering of the yield strength of the material as a result of increase in temperature. In the case of steel, serious deterioration in strength begins at temperatures of around 500 °C. The most vulnerable parts of the structure are compression elements and the compression flanges of beams.

Most regulatory authorities specify minimum requirements for performance of a building in fire depending on the type of occupancy for which the building is intended, the size of the building, and the extent to which it is compartmentalised. So far as the structure is concerned the performance is measured in terms of a minimum period of fire resistance (from 30 minutes to 4 hours, depending on the size of the building and the type of occupancy). Roof structures and single-storey structures are normally exempted.

In the case of steel structures, two types of strategy may be employed to meet the fire performance requirements, namely the 'fire protection' strategy and the 'fire resistance' strategy. 'Fire protection' is by far the more common. This involves the encasing of the structure in a layer of insulating material in order to reduce the rate at which the temperature of the steel increases, so that the critical temperature is not reached within the required fire-rating period.

Where the strategy is one of fire resistance the objective is to minimise or entirely eliminate the need for protection. To comply with fire regulations by this method the designer of a structure is required to demonstrate, by calculation or some other simulation method, that the structure will not lose its integrity or become unstable in a fire before the required fire-rating time has elapsed. Various strategies can be adopted to achieve this. For example, the structure can be designed such that the actual stresses under service loads are maintained at a low level in parts of the building which are vulnerable to fire (e.g. columns). Bearing in mind that the problem with fire is the reduction in yield strength due to temperature increase, and that instability will not occur until the yield strength becomes reduced to the level of the actual stress in an element, the lower the stress in the material the longer will it take for that point to be reached following exposure to fire. If this method is applied, therefore, the need for protection of steelwork by insulating material can theoretically be eliminated or the amount of the required protection reduced. The problem with the fire resistance strategy is the lack of data on which to base the calculations required to prove that the necessary resistance time will be achieved. This is currently an active field of research, however, and the situation may be expected to improve. The protection of steelwork ('fire protection') is, however, the strategy which is normally adopted at present.

Three types of insulating material are used to protect steel structures. The first group comprises the traditional materials of concrete, masonry and insulating board, such as plasterboard. Concrete is perhaps the most effective but is expensive and can add significant dead load to the structure. Insulating boards are also expensive because the fixing of them is labour intensive. The second type of system consists of spray-on material based on rock-fibre or vermiculite. This is currently the cheapest method and is particularly convenient for complex shapes or connections but it is generally regarded as unsightly and only suitable in situations in which the structure will not be exposed to view, such as above false ceilings. A third type of product is the intumescent coating. With this system the protecting material is very thin (treated steelwork has the appearance of having been painted) but on exposure to fire, a chemical reaction occurs which generates gas and causes the material to become a foam and

form an insulating layer. Intumescent coatings are particularly effective in situations in which the steelwork forms part of the architectural language of the building.

A further fire protection strategy is the use of water cooling as a means of maintaining the temperature of a steel structure at an acceptable level. This strategy normally requires that hollow-section elements be adopted and in practice has only very rarely been utilised. A prominent example is the Centre Pompidou in Paris in which the circular cross-section columns are protected in this way. Even in this building, however, conventional fire protection systems were used for all other structural elements.

3.6 Structural forms

3.6.1 Introduction
Steel frames are assemblies of structural elements which are selected or built up from the standard range of components. They are prefabricated structures whose constituents are prepared to a semi-finished state in the factory or workshop before being transported to the site for erection.

The design of the connections between the elements is an important aspect of the planning of steel frames. Two types of fastening element are used in steelwork: bolts and welding. In most cases, welding provides a better structural junction than bolts, but bolts are easier to connect on site. Most steel frames are therefore designed so that they can be prefabricated by welding into parts which are small enough to be transported easily (not more than 10 m in length) and these are then connected together on site by bolting; most steel frames therefore have joints in which a combination of welding and bolting is used.

So far as the structural performance is concerned there are two basic types of joint, hinge-type joints and rigid joints. Hinge-type joints transmit shear and axial force between elements but not bending moment. They are, in other words, incapable of preventing elements from rotating relative to one another

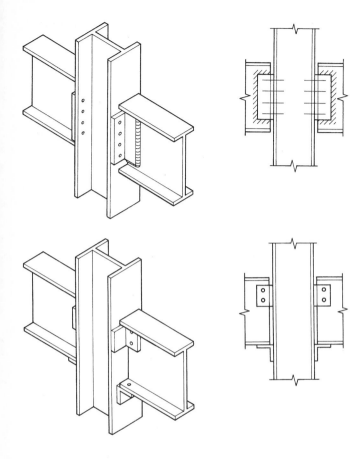

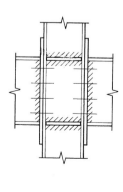

and therefore behave in a similar way to a hinge. Rigid joints transmit shear and axial force and bending moment between elements and therefore provide a fully fixed connection. Both types of joint can be made either by welding or by bolting but are normally made by a combination of both; rigid joints are more complex and more expensive to design and to construct. Examples of both types of joint are shown in Figs 3.23 and 3.24.

The selection of the joint types and their disposition in the frame is an important consideration in the design of a steel structure. The principal factors which are affected by the joints are the stability of the structure and its state of statical determinacy.[10] Frames in which the majority of joints are of the hinge type (simple frames) are usually statically determinate and are easier and cheaper to design and construct than those with rigid joints, which are normally statically indeterminate. Frames with hinge-type joints are usually unstable in their basic form and require separate bracing systems for stability. Rigidly jointed frames, due to their statical indeterminacy, are more efficient types of structure and therefore have more slender elements than hinge-jointed equivalents; they are also self-bracing.

Fig. 3.23 Hinge-type joints. The joints illustrated here are capable of transmitting shear force but not bending moment and are therefore hinge-type connections. Note that both welding and bolting are used. Welding is more efficient in transmitting load but bolting is easier on the building site. The joints are detailed so that the site-made part of the connection is made by bolting.

10 The question of statical determinacy versus statical indeterminacy is an important consideration in structural design because it affects a number of aspects of the performance of structures. For a discussion of the issues involved see Appendix 3 of Macdonald, *Structure and Architecture*.

Fig. 3.24 Rigid joints. These joints allow the transmission of both shear force and bending moment between elements. Note that they too are executed by a combination of bolting and welding.

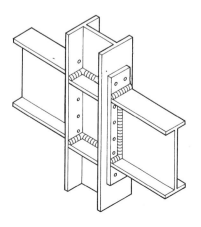

Steel frames can therefore be thought of as either 'simple' frames, that is frames in which the majority of the joints are of the hinge type, or as 'rigid' frames, in which case the joints are rigid. Most frames are of the 'simple' type and therefore require bracing systems for stability (see Sections 3.6.2 and 3.6.3).

As with other types of structure, the configuration which is adopted for a steel frame is influenced by the principal types of load which it will carry and, in particular, by the characteristics of whichever of the applied loads is dominant. In the descriptions which follow frames are subdivided into three categories depending on the dominant type of applied load. The three main types of applied load on architectural structures are imposed roof load, imposed floor load and wind load. The three resulting categories are single-storey frames, in which imposed roof loading is the dominant load, low- and medium-rise multi-storey frames, in which imposed floor load is the dominant form of load, and high-rise multi-storey frames, in which wind load is a significant structural factor in the design. A very large number of different geometries and configurations is possible within each category. Only the most basic forms are actually described here.

3.6.2 Single-storey frames

3.6.2.1 Introduction
The variety of different possible arrangements for single-storey frames is very large and they are placed here into the two broad categories of one-way- and two-way-spanning systems. Each of these categories is then further subdivided.

3.6.2.2 One-way-spanning systems
There are two basic types of one-way-spanning single-storey frame, the main difference between them being that in one the principal elements are spaced close together and carry the roof cladding directly, and in the other they are located at a fairly wide spacing and are linked by a secondary system of elements to which the cladding is attached. The second of these arrangements is more versatile and the

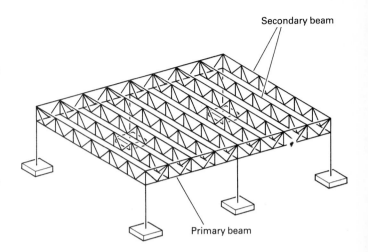

Secondary beam

Primary beam

Fig. 3.25 This is a typical layout for a single-storey frame in which lightweight steel elements are used. In the primary–secondary beam system illustrated, closely spaced triangulated joists are arranged parallel to one another and are supported by primary beams, which are in turn supported by columns. The column grid is rectangular with the secondary beams running parallel to the long side of the rectangle.

majority of single-storey steel frames are in this category.

Figure 3.25 shows a typical arrangement of elements in the first type of frame. It is normal for only every third or fourth element to be supported directly on a column with the remaining elements being carried on beams running at right angles to the principal direction to produce a rectangular column grid. The roof geometry can be given a wide variety of elevational forms including flat, mono-pitch, duo-pitch and curved, but a regular-plan geometry consisting of equally spaced parallel elements is normally adhered to.

The principal elements in this type of frame carry relatively small areas of roof and are therefore lightly loaded. This favours the use of light, efficient types of structural element such as proprietary lightweight lattice joists, based on cold-formed sections (Figs 3.20 and 3.26). 'Improved' hot-rolled sections, such as castellated beams (Figs 3.16 and 3.27), are also suitable but the use of standard hot-rolled sections such as the universal beam is rarely economic.

Fig. 3.26 A single-storey steel framework is shown here comprising lightweight Metsec joists based on cold-formed sections. [Photo: Photo-Mayo Ltd]

Fig. 3.27 Castellated beams are arranged here in one of the standard configurations for single-storey steel frameworks. [Photo: Pat Hunt]

Table 3.4 Span ranges and typical element sizes for lightweight steel joists in single-storey frame structures

Secondary beam span, L (m)	Secondary beam spacing, P (m)	Secondary beam depth (mm)	Column spacing, S (m)	Column width (mm)	Column height (m)
4	1.0	200	3	50	3.0
5	1.0	200	3	60	3.0
6	1.5	250	4.5	70	3.5
7	1.5	300	4.5	70	3.5
8	2.0	350	6	90	4.0
9	2.0	350	6	90	4.0
10	2.0	400	6	90	4.0
12	2.5	550	9	100	5.0
14	2.5	600	10	150	5.0
16	3.0	700	12	150	5.0
18	3.0	800	12	150	5.0
20	3.0	900	15	180	5.0
22	3.0	1100	15	180	6.0
24	3.0	1300	18	200	6.0
26	3.0	1500	18	250	6.0
28	3.0	1600	24	250	6.0
30	3.0	1800	24	250	6.0

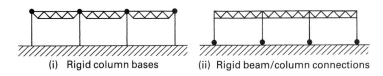

(i) Rigid column bases   (ii) Rigid beam/column connections

Fig. 3.28 Single-storey frameworks based on lightweight joists can be braced by the rigid-joint technique. The rigid joints can be either at the bases or at the tops of the columns. All columns must normally be braced, however.

The size of the column grid depends on the type of structural element which is specified; the normal range is from approximately 6 m × 6 m to 10 m × 25 m with element spacings of around 1.5 m to 2 m. The columns are usually hot-rolled sections, either universal columns or rectangular hollow sections. Approximate sizes for the normal span range of this type of structure are given in Table 3.4.

Both self-bracing and braced versions of this configuration are possible.[11] Self-bracing frames have either rigid column base connections or rigid beam/column joints (Fig. 3.28).

Where hinge-type connections are used, bracing in both horizontal and vertical planes is required and diagonal bracing can be used for both (Fig. 3.29). Diaphragm bracing is an alternative, however. In the vertical plane this can be provided by masonry walls, if these are adequately tied to the columns, and in the horizontal plane the roof decking can perform this function, where it has sufficient strength.

Figure 3.30 shows a typical plan layout, with typical span ranges, for a single-storey frame in which the primary elements are arranged at fairly wide spacings and a secondary structure, on which the roof cladding is mounted, is provided to link them together. The primary elements must be stronger than in the previous type of frame and hot-rolled sections are

11 For a more detailed explanation of the bracing requirements of frameworks see Macdonald, *Structure and Architecture*, Section 2.2.

77

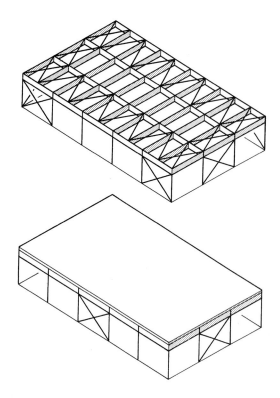

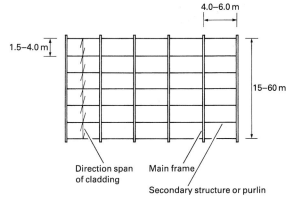

4.0–6.0 m

1.5–4.0 m

15–60 m

Direction span
of cladding

Main frame

Secondary structure or purlin

Fig. 3.30 Basic plan arrangement for a single-storey steel framework in which the primary elements are strong girders based on hot-rolled sections.

Fig. 3.29 Diagonal bracing schemes for single-storey frameworks with lightweight elements.
(a) Diagonal bracing in both horizontal and vertical planes.
(b) Diagonal bracing in the vertical planes; roof cladding used as diaphragm bracing in the horizontal plane.

normally used. A wide range of element types can be adopted depending on the span involved (Fig. 3.31). For medium spans, universal or castellated beams are feasible for the principal elements and if the portal frame configuration is adopted a very wide span range is possible with these (15 m to 60 m). Lattice girders of various geometries, including plane and space trusses, in which the individual sub-elements have hot-rolled sections (rectangular or circular hollow sections, angles, channels, etc.), are also used over a wide range of spans. The spacing of the primary elements must usually be restricted to a maximum of around 6 m to 8 m, regardless of the primary span, so that the secondary elements are not excessively large. As in the case of frames with closely spaced primary elements, a wide range of

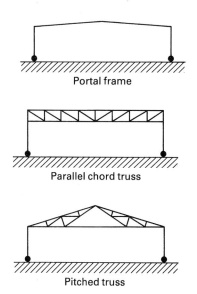

Portal frame

Parallel chord truss

Pitched truss

Fig. 3.31 Typical examples of primary elements in hot-rolled steelwork.

Table 3.5 Approximate depths and span ranges for trusses of hot-rolled sections in single-storey steel frames

Span (m)	Depth of main frame (mm)		
	Solid web	Plane truss	Space truss
10	450	1000	1000
15	600	1200	1200
20	700	1400	1400
30	900	1800	1600
40	1200	2500	2200
50	–	3000	2800
60	–	4000	3800
70	–	5000	4800
80	–	6000	5500
100	–	8000	6000

Table 3.6 Approximate span ranges and element dimensions for steel portal frameworks

Span, L (m)	Spacing, S (m)	Height, H (m)	Main Frame	
			Rafter depth (mm)	Depth at knee (mm)
10	3	6	300	600
15	4	7		
20	5	8	450	800
25	6	9		
30	7	10	550	1050
35	8	11		
40	9	12	600	1200
45	10	13		
50	10	14	700	1350
55	12	15		
60	12	16	750	1600

elevational roof geometries can be adopted but a regular plan layout must normally be adhered to. Table 3.5 gives an indication of the depths required for the primary elements in parallel chord arrangements. Approximate sizes for portal frameworks are given in Table 3.6.

Frames of this kind are constructed with both rigid and hinge-type joints and the bracing requirements are dependent on the particular configuration of joints which is adopted. A common arrangement is illustrated in Fig. 3.32.

3.6.2.3 Two-way-spanning systems
In two-way-spanning systems primary elements are provided in two principal span directions. The structure can take the form of intersecting sets of single-plane elements, such as triangulated trusses (Fig. 3.33), or of a fully three-dimensional triangulated space framework (Fig. 3.34).

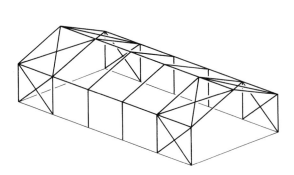 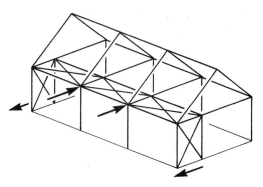

Fig. 3.32 Bracing of single-storey frameworks, based on strong, hot-rolled primary elements.
(a) This is a typical bracing arrangement for a portal-frame structure. The frame is self-bracing, by rigid joints, in the across-building direction but triangulated bracing girders are required in the vertical and roof planes adjacent to the ends of the building to guarantee stability in the along-building direction.
(b) Where a framework is not self-bracing in the across-building direction an arrangement of bracing girders must be provided in both horizontal and vertical planes.

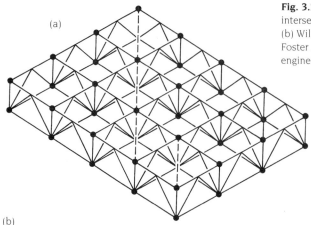

(a)

Fig. 3.33 (a) Two-way-spanning roof structure formed by intersecting plane frameworks.
(b) Willis, Faber & Dumas office, Ipswich, England, 1974. Foster Associates, architects; Anthony Hunt, structural engineers. (Photo: Pat Hunt).

(b)

In the first type of system the elements may intersect at right angles or obliquely. The oblique grid pattern, which is called a 'diagrid' (Fig. 3.35), is the more efficient configuration because the girders which span across the corners are relatively short and can provide a measure of support for the longer spanning girders which run between the corners.

True space frames, that is fully three-dimensional structures whose fundamental units are pyramids or tetrahedra (Fig. 3.34), allow a more efficient use to be made of material than in the case of the intersecting plane-frame type because the disposition of the elements produces a more satisfactory distribution of internal forces. True space frameworks also can

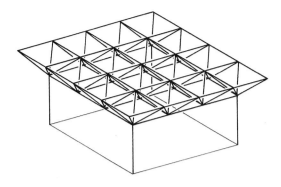

Fig. 3.34 Fully three-dimensional two-way-spanning space framework.

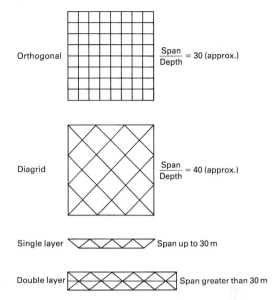

Fig. 3.35 Plan and cross-sectional arrangements for space frameworks.

be given the rectangular or diagrid form depending on the orientation of the elements with respect to the principal span directions.

Space frameworks of both of the kinds described above are highly statically indeterminate and this, together with the fact that they are triangulated, allows very efficient use of structural material. They are, however, complicated to design and construct and are therefore more expensive than the one-way-spanning systems described above. Truly three-dimensional frames are economically competitive with conventional structures at spans greater than 20 m; space frames which consist of intersecting plane frames are used mainly in the span range 15 m to 20 m.

Because space frames are two-way-spanning systems, the most suitable overall plan-form is a square in which the frame is supported around its entire perimeter on regularly spaced columns or walls (Figs 3.36 and 3.37). A square plan-form with supports at the corners is another common arrangement. The high degree of statical indeterminacy allows an irregular pattern of support to be used, however, if this

Fig. 3.36 Support arrangements for space frameworks. Space frameworks are normally supported either by regularly spaced perimeter columns around the entire perimeter or by a symmetrical arrangement of columns or piers set back from the perimeter. The two-way-spanning capability of the space frame allows, however, an irregular pattern of support. This is sometimes the reason for adopting a space frame as the horizontal structure in a building.

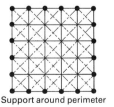

Support around perimeter

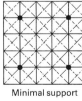

Minimal support

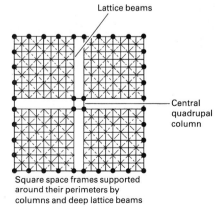

Lattice beams

Central quadrupal column

Square space frames supported around their perimeters by columns and deep lattice beams

Fig. 3.37 A Mero space frame, supported around its perimeter, acts as the roof structure in this single-storey building. Note the vertical-plane diagonal bracing elements which must be provided in each external wall of the building. The fully triangulated space framework acts as horizontal-plane bracing. [Photo: BICC]

is required, and space frames can even be given an irregular overall form so long as the internal geometry is fully triangulated.

One of the advantages of the space frame is that the individual elements are small and light. The parts can therefore be easily transported to the site and no large equipment is required for erection; often the structure can be assembled at ground level, by bolting or welding, and subsequently lifted by a jacking process into its final position. This greatly simplifies the construction.

The principal disadvantage of the space frame is its complexity. In particular, the high degree of geometric complexity makes difficult the design and manufacture of the joints, and

the high degree of statical indeterminacy aggravates this problem by requiring that components be manufactured to small tolerances so that the 'lack-of-fit'[12] is minimised.

The difficulties which are associated with construction can be reduced if a high degree of standardisation is adopted and the space frame is a type of structure which can be most economically produced in the form of a proprietary system. A number of these are currently available and they exploit the potentially high structural efficiency of the space-frame configuration while at the same time allowing a reasonable level of cost to be achieved through the standardisation of components and by means of the economies of production-line manufacture (Figs 3.38 to 3.41). Standardisation reduces the structural

12 See Macdonald, *Structure and Architecture*, Appendix 3 for an explanation of the 'lack-of-fit' problem.

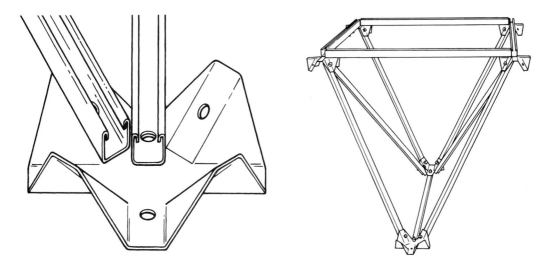

Fig. 3.38 The elements of the Unistrut space deck system. This is based on lightweight cold-formed sections and an ingenious jointing component. The efficiency with which load is transmitted through the joints is low, however, and the system is capable of relatively short spans only.

efficiency, but proprietary space-frame systems are nevertheless normally significantly lighter than conventional single-plane, one-way-spanning arrangements. They are also cheaper than one-off space-frame systems and the precision which can be achieved with economy by the use of production-line techniques in fact makes the proprietary space frame one of the cheapest of the more sophisticated forms of structure.

Fig. 3.39 Spacedeck space frame system supported around its perimeter at the Aberfan Centre, Wales. [Photo: Henry Morgan]

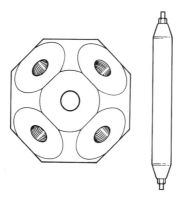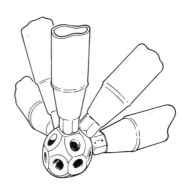

Fig. 3.40 (a) Basic components of the Mero space frame. This highly versatile system allows great freedom to the building designer because the lengths of the elements are not predetermined. It is possible to use this system to create a wide variety of structural forms including domes and vaults as well as flat space decks.
(b) Church, Milton Keynes, England, with Mero spacedeck roof. [Photo BICC]

Standardisation normally imposes fairly severe geometric constraints on the design, however, both on the overall form which can be adopted and in the positioning and spacing of the supports. Some proprietary systems can be used for flat decks only (Figs 3.38 and 3.39) but others are more versatile and can have more complicated geometries (Figs 3.40 and 3.41). Most proprietary systems are designed for the short-span range (to cover areas of

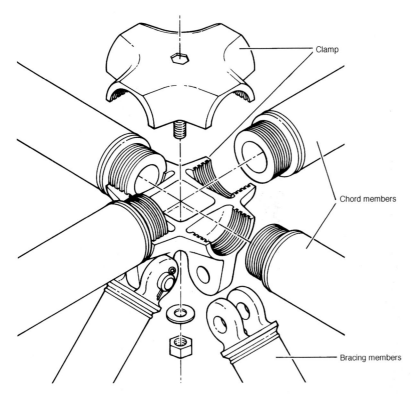

Clamp

Chord members

Bracing members

Fig. 3.41 Nodus joint. The Nodus joint is designed for use with hot-rolled steel hollow sections. It is basically a clamping device which is used in conjunction with end pieces which are welded to standard hollow sections. It allows a wide range of space frame geometries to be constructed and gives the designer freedom, not only to select different frame configurations, but also to vary the sizes of the constituent elements.

around 10 m by 10 m to 30 m by 30 m on plan). Requirements for larger spans, or for more complex geometries, can frequently only be satisfied by use of a one-off, specially designed system. Such systems are usually all-welded structures with hot-rolled hollow section elements. An indication of the span capabilities and depths required for the systems illustrated in Figs 3.38, 3.40 and 3.41 is given in Tables 3.7, 3.8 and 3.9.

Table 3.8 Approximate depths and span ranges for the Mero space framework

Span (m)	Module (m)	Frame depth (m)
up to 15	2 to 3	up to 1.5
15 to 27.5	2.4 to 3	1.5 to 2.1
27.5 to 36	2.4 to 3.6	2.1 to 2.5
36 to 50	3.6 to 4.8	2.5 to 4.0
50 to 100	4.8 to 6.0	3.6 to 4.8

Table 3.7 Dimensions and span ranges for the Unistrut space framework

Module (m)	Depth (m)	Span range (m)	
1.22	1.0	8.5 to 15.2	Column support at corners
		8.5 to 31.7	Support around perimeter
1.52	1.26	7.6 to 18.2	Column support at corners
		7.6 to 36.6	Support around perimeter

Table 3.9 Approximate depths and maximum span ranges for the Nodus space framework

Module	2 to 3 m	
Frame depth	$\dfrac{span}{20}$	Column support at corners
	$\dfrac{span}{15}$	Support around perimeter
Maximum span	50 m	Column support at corners
	60 m	Support around perimeter

Fig. 3.42 Terminal building at Stuttgart Airport, Germany. The roof structure here consists of a series of complex 'trees', constructed from steel hollow sections, which support a regular grid of secondary elements on which the cladding is mounted. The sub-elements of the trees are subjected to high levels of bending load and it is the great strength of steel which makes this type of arrangement possible. [Photo: A. Macdonald]

3.6.2.4 Frames with special geometries

The overwhelming majority of single-storey steel frameworks for buildings can be placed into one or other of the categories described above; these represent the most sensible, straightforward and economic ways of using the material. There are, of course, exceptions which arise due to the existence of unusual design requirements or simply as a response to the desire to produce an unusual or spectacular structure.

One example of a departure from the normal frame arrangement is a two-way-spanning system based on a series of structural 'trees'.

The basic elements of this type of structure consist of a column unit (the 'trunk') which has a rigid joint at its base, and a series of cantilevering horizontal members (the 'branches'). In most cases this primary structure supports a secondary system of parallel beams to which the roof cladding is attached. The plan of each 'tree' unit can have any shape but is usually square with plan dimensions in the range 15 to 40 m. A number of units are placed alongside one another to give a building of the required size. The advantage of the system is that each unit is entirely independent structurally from its neighbours. This situation allows for simple extension or alteration of the building, which is frequently the reason for its adoption. This is also a form of construction which lends itself to very fast erection and the rapid production of a weathertight envelope.

Two notable buildings which have been constructed in this way in recent years are the terminals at Stuttgart (Germany, Fig. 3.42) and Stanstead (England) Airports. The Renault Warehouse building at Swindon, England, by Foster Associates (Fig. 2.6), and the Fleetguard

Fig. 3.43 Inmos Microprocessor Factory, Newport, Wales, 1982. Richard Rogers Partnership, architects; Anthony Hunt Associates, structural engineers. A mast-and-tie arrangement is used here to achieve a long span with a one-way-spanning structural system. [Photo: Alastair Hunter]

Factory at Quimper, France, have structural arrangements which are a combination of the structural tree system and the continuous two-way-spanning frame. In each of these cases the structural trees are not independent units.

In the Fleetguard Factory the sizes of the horizontal elements were kept low by the use of mast-and-tie systems. Mast-and-tie systems can also be used in the context of one-way-spanning frame arrangements. Examples of this are the Inmos Factory, England by Richard Rogers (Fig. 3.43) and the Ice Rink at Oxford, England by Nicholas Grimshaw (Fig. 3.44). In all of these cases the tie rods or cables serve to provide a regular pattern of vertical support for the horizontal elements and may be regarded as substitutes for columns at these

Fig. 3.44 Ice Rink, Oxford, 1984. Nicholas Grimshaw & Partners, architects; Ove Arup & Partners, structural engineers. The mast-and-tie system was used here to achieve a large column-free interior. The prevailing ground conditions were a factor in the selection of this particular structural configuration. [Photo: Ove Arup & Partners]

points. The justification for the elaborate mast-and-tie system is normally that it allows a large column-free interior to be achieved, either because this is a space-planning requirement (as was the case at the Oxford Ice Rink) or because great flexibility is needed in the planning of the interior arrangements of the building.

3.6.3 Low- and medium-rise multi-storey frames

3.6.3.1 Introduction
In low- and medium-rise multi-storey frames, that is frames with up to 30 storeys, the principal structural factor which affects the plan arrangement is the need to provide a suitable system of beams and columns to support the gravitational load on the floors. The need for the structure to have adequate lateral strength to resist wind load is, of course, also an important consideration, particularly with the higher structures.

3.6.3.2 The floor grid
In most multi-storey frames the floors are of reinforced concrete. A number of different types of floor slab are used including *in situ* or precast concrete slabs with simple rectangular cross-sections, voided precast slabs of various kinds, and *in situ* slabs which are cast on a permanent formwork of either steel or precast concrete (Fig. 3.45). The concrete slabs are normally made to act compositely with the floor beams to improve the overall efficiency of the structure (Fig. 3.46). The floor slabs are almost invariably of the one-way-spanning type and must be supported on a parallel system of beams spaced 2 m to 6 m apart depending on the slab type.

The column grid can take various forms depending on the space-planning requirements of the building. A very common basic

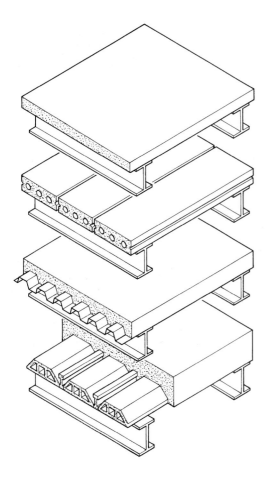

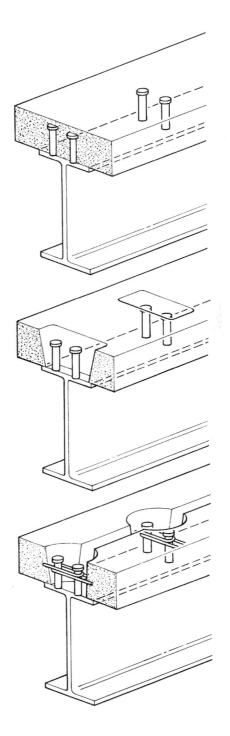

Fig. 3.45 Typical floor-slab systems for steel frameworks.
(a) In situ reinforced concrete flat slab.
(b) Precast concrete floor units.
(c) In situ concrete on profiled steel permanent formwork.
(d) Composite precast and in situ concrete.
All of these systems are normally one-way-spanning and require to be supported on a parallel arrangement of steel beams. System (a) can be two-way-spanning.

arrangement is one in which the columns are positioned at every third or fourth beam and intermediate beams are carried on a second system of beams which run at right angles to the direction of the main beams (Fig. 3.47). This produces a column grid which is rectangular. A variation of this is the grid in which columns are provided at every main beam where these meet the perimeter of the building, and at every third or fourth beam in the building's interior (Fig. 3.48). This eliminates

Fig. 3.46 Shear studs welded to the top flanges of steel floor beams allow composite action to be developed between the beams and the floor slabs. This reduces the size required for the beam. As is shown, such action is possible with both in situ and precast floor systems.

89

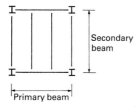

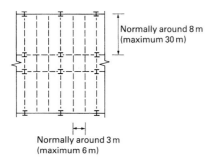

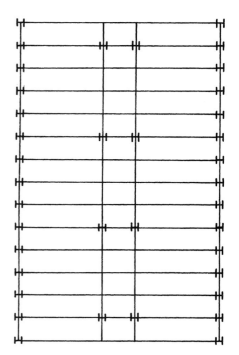

Fig. 3.47 Typical layout of floor beams in a steel framework. (a) The basic unit consists of a series of secondary beams in a parallel arrangement at equal spacing. The spacing depends on the span capability of the floor-slab system. The secondary beams are used in conjunction with primary beams to give a column grid which is more-or-less square. (b) Variations to the basic arrangement are used to accommodate space-planning requirements.

Fig. 3.48 A common variation on the basic floor grid arrangement is the provision of columns for each secondary beam at the building' perimeter. The closely spaced columns eliminate the need for a separate cladding support system.

the need for a secondary system of mullions in the external wall on which to mount the cladding.

The types of steel sections which are used for the elements in the floor grids depend on the span and on the intensity of the load. For grids in the span range of 4 m × 6 m to 6 m × 12 m, carrying office-type loading, universal beam or castellated beam sections are normally adopted; an indication of the depths of beam which are required is given in Table 3.10.

For economy in construction and simplicity in the detailing of the structure, standardisation of elements is desirable. This is facilitated if the spans of the principal elements are kept constant. Where beams with different strengths are required, a constant floor depth is sometimes achieved without compromising efficiency by the use of different weights of beam within a given serial size.

Where a larger column grid is required than is possible with standard sections, that is a grid with dimensions in excess of around 20 m, various forms of built-up sections can be used to provide stronger beams. Plate girders and compound beams allow the span range to be extended but for very large spans lattice girders are required. In cases where very long spans must be achieved it is frequently advantageous to maintain a basic floor grid, based on standard sections, and use a tertiary system of girders to achieve the long span (Fig. 3.49).

The depth of long-span structures tends to be large and various measures can be adopted to minimise the amount of unusable volume

Table 3.10 Typical element sizes in multi-storey steel frameworks

| Floor beam span (m) | Floor beam depth (mm) | | | Column cross-section (H-section or square hollow section) (mm) | | | |
| | I-section | I-section in composite structure | Castellated | Storey height (m) | | | |
				3	4	5	6
4	250	200		150	150	150	150
6	400	380	550	150	200	200	200
8	530	450	600	200	200	200	250
10	760	600	900	200	250	250	250
12	900	760	1150	250	250	250	250
14	900	830	1370	250	250	300	300

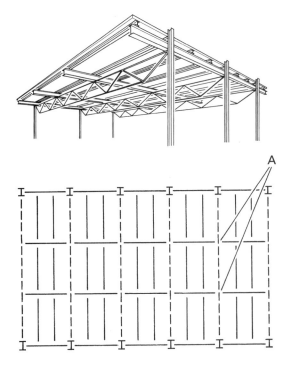

Fig. 3.49 Where long spans are required in multi-storey frameworks it is normal to provide a tertiary system of beams. In the example illustrated, triangulated girders, spanning the entire width of the building (shown dotted on the plan), are substituted for the columns at points A. [after Hart, Henn and Sontag]

which is created within the building. Where very deep lattice girders are involved, for example, the depth of these can sometimes be made the same as the storey height of the building so that the girders can be incorporated into walls, as in the staggered-truss and interstitial-truss systems (Fig. 3.50).

3.6.3.3 Bracing systems for medium-rise, multi-storey frames

As in the case of single-storey frames multi-storey frames can be of either the 'simple' or the rigid type. Where the joints are rigid, the frame is self-bracing and the need to provide lateral stability is not a major factor in the determination of the overall form. Where hinge-type joints are used it is normal practice for all beam/column connections to be of the hinge type and for the columns to be jointed at every second storey level, also with hinge joints (Fig. 3.51). This fully hinged configuration has a number of advantages: it simplifies both the analysis and the erection of the structure, and it also allows the accommodation of thermal expansion and minor foundation movements without the introduction of stress into the steelwork. In its basic form this type of frame is highly unstable, however, and additional bracing must be provided.

To render the 'simple' type of frame stable, a limited amount of vertical-plane bracing, of the diagonal or diaphragm types, must be incorporated into the arrangement in two orthogonal (mutually perpendicular) directions and this must be linked to all other parts by horizontal-plane bracing at every level. The action of a multi-storey frame in response to wind loading is shown diagrammatically in Fig. 3.52.

Fig. 3.50 Where long spans require the use of deep tri-angulated girders this can greatly increase the structural depth of the floors. The two systems shown here allow deep girders to be accommodated without an increase in storey height. The penalty which is paid for this is that it places restrictions on space-planning.

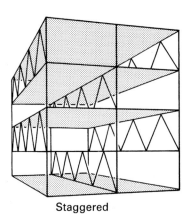 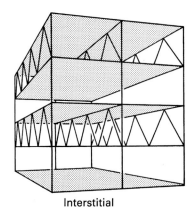

Staggered   Interstitial

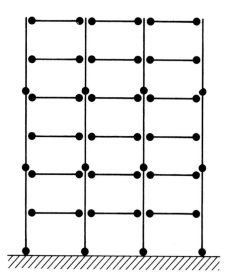

Fig. 3.51 Typical arrangement of joints in a hinge-jointed multi-storey framework.

The disposition of the vertical-plane bracing is a factor which affects the general planning of the building. It is normally positioned in as symmetrical an arrangement as possible and convenient locations for diagonal bracing are the perimeter walls or around stair wells, lift wells and service cores (Fig. 3.53). If diaphragm bracing is used it is frequently provided in the form of self-supporting cores of masonry or reinforced concrete (Fig. 3.54), which are normally constructed in advance of the erection of the steelwork and which are used to carry vertical as well as horizontal loads.

The reinforced concrete floors of the structure are usually capable of acting as horizontal-plane bracing and the need for this does not usually affect the general planning of the building. Where a lightweight roof cladding is used, a horizontal-plane girder (normally of the triangulated type) is provided at the topmost level but this can be accommodated within the roof structure and does not usually present a planning problem.

To sum up, two factors must be considered in the planning of low- or medium-rise multi-storey frames. These are firstly, the geometry of the floor grid and its relationship to the column grid and secondly, the provision of bracing. A general arrangement must be devised which both produces a satisfactory structure and allows the space-planning requirements of the building to be met.

3.6.3.4 Multi-storey frames with special geometries

Although most of the steel frames which are constructed in practice have characteristics which are similar to those which are described above, the need for a frame which has quite different characteristics does sometimes arise.

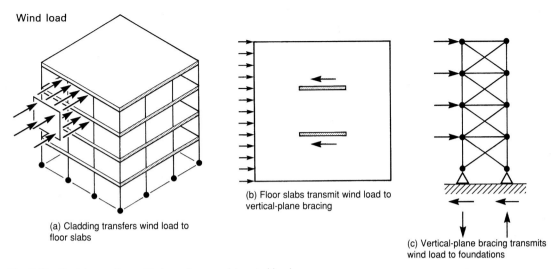

Wind load

(a) Cladding transfers wind load to floor slabs

(b) Floor slabs transmit wind load to vertical-plane bracing

(c) Vertical-plane bracing transmits wind load to foundations

Fig. 3.52 Resistance of a multi-storey framework to wind load.
(a) The wind loading acts on the cladding and is transmitted by the cladding support system to the floors of the building.
(b) In this plan view the wind loading appears as a distributed load on the edge of the floor slab, which is restrained by the vertical-plane bracing.
(c) The vertical-plane bracing is shown here independently of the rest of the frame. The loads received from each floor are indicated. These are transmitted to the foundations by the bracing girder (which consists of the columns forming the bay, together with the diagonal bracing elements).

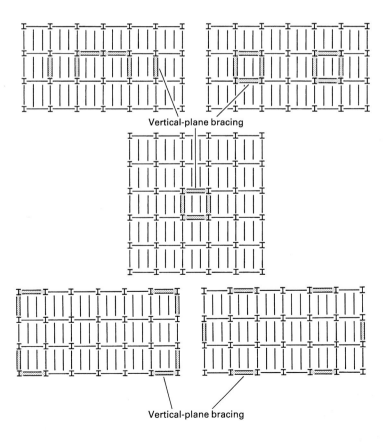

Vertical-plane bracing

Vertical-plane bracing

Fig. 3.53 The locations of vertical-plane bracing must be compatible with space-planning requirements. Typical arrangements, in which vertical-plane bracing is located in the perimeter walls or around service cores, are shown here.

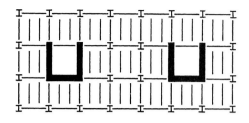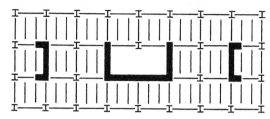

Fig. 3.54 Where diaphragm-type vertical-plane bracing is used it is normally constructed in either masonry or reinforced concrete. It can consist simply of infill walls in the steel framework or, as is shown here, it can take the form of free-standing cores to which the steelwork is attached. In the latter case the cores are normally constructed to the full height of the building in advance of the steelwork being erected. As these cores normally contain stairs they provide access to all parts of the building throughout the construction sequence.

An example of a special requirement is the situation in which the lowest storey of a building must have a different plan geometry from upper floors because it contains a radically different type of accommodation. This can create the need for a special frame, especially if, as is often the case, there is a requirement to reduce the number of columns in the lowest storey. One way in which this is achieved is by locating a number of very large girders at first-floor level so as to transmit the loads from the columns in the upper storeys to a smaller number of columns in the ground floor. Another arrangement which has been used is the suspended frame (Fig. 3.55). This has the additional advantage that it reduces the size of the majority of the vertical elements because tension elements have a smaller cross-section than equivalent columns. The adoption of this type of system therefore increases the usable floor area per storey.

3.6.3.5 High-rise multi-storey frames
The dominating factor in the design of high-rise frames is the need to provide sufficient lateral strength and rigidity to resist wind load effectively. As in the case of low- and medium-rise multi-storey buildings, the floor must be designed to provide effective resistance to gravitational loads and to act as a horizontal

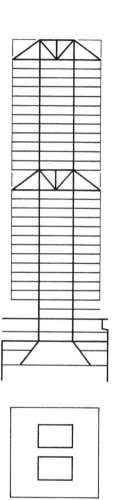

Fig. 3.55 Suspended frame arrangement. The floors here are suspended from one or other of two massive trusses which are supported on a central core. The total volume of vertical structure is theoretically lower than with the traditional column grid arrangement because the concentration of the compressive function into one massive central element allows higher compressive stresses to be used (because the slenderness of the single compressive element is relatively low). The structure of the HongkongBank Headquarters (Fig. 1.3) is based on this principle.

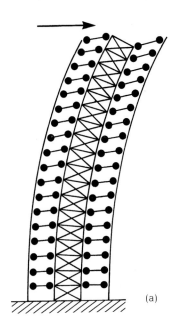

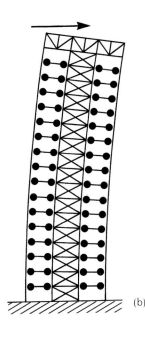

(a)

(b)

Fig. 3.56 Belted truss system.
(a) The diagonally braced core is here carrying all of the horizontal load.
(b) The introduction of the rigid girder at the top of the building allows the perimeter columns to act compositely with the core to resist the horizontal load.

diaphragm to conduct wind forces to the verti-cal-plane bracing. In the case of high-rise structures more attention is given to the latter function but the floor grids which are adopted are nevertheless similar to those which are used in lower structures.[13]

For frames with more than 30 storeys, neither rigid-frame action nor diagonal or diaphragm core-bracing are capable of provid-ing sufficient lateral strength in the vertical plane to resist wind load effectively. A combin-ation of core diaphragm bracing and rigid joints provides sufficient lateral strength for buildings of up to 60 storeys in height; the belted truss system (Fig. 3.56), which allows the core to act compositely with the perimeter columns of the building and which reduces the bending moment on the core due to the lateral loads, is another arrangement which is suitable for buildings in this height range. For higher buildings the lateral strength is increased by making use of the full cross-sectional width of the building, so that it behaves like a single vertical cantilever.

Various techniques have been used to achieve this. In the framed-tube system (Fig. 3.57), which was used in the World Trade Centre building in New York, the closely-spaced perimeter columns form a cantilever tube; interior columns play no part in the resistance of lateral load in this system and the whole of the wind load is resisted by the external walls. The walls which are parallel to the direction of the wind are rigid frames and form a shear connection between the walls which are normal (at right angles) to the wind direction, which act as flanges. The building is therefore a box girder which behaves as a vertical cantilever in response to lateral load. The action of the trussed tube (Fig. 3.58) is similar. In the case of the tube-in-tube structure (Fig. 3.59) the stiffness of the tube, which is formed by the perimeter walls, is augmented by composite action with a core tube. This requires that floor structures be stiff enough to provide a shear connection between the two tubes. Bundled-tube structures (Fig. 3.60), which were developed to reduce the shear-lag effect which occurs in single-tube construction, produce the stiffest buildings for a given size of cross-section.

13 See, Schueller, *High Rise Building Structures*, John Wiley, London, 1977.

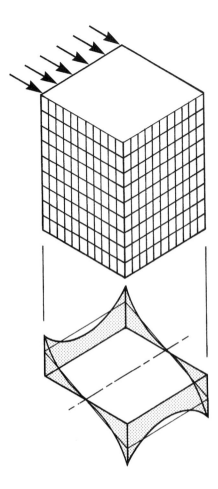

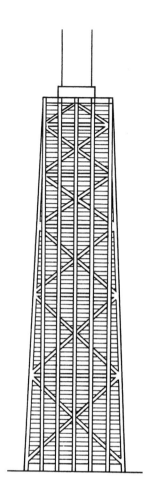

Fig. 3.57 In the framed-tube system the columns on the perimeter of the building are closely spaced. The building acts as a vertical cantilever in response to wind loading. The cross-section of the cantilever is a rectangular tube formed by the closely spaced perimeter columns. Those on the windward and leeward faces of the building act as flanges while the walls which are parallel to the wind provide a shear connection between these by rigid frame action. The World Trade Centre buildings in New York are perhaps the most well-known buildings to be based on this principle. The lower diagram here gives an indication of the variation in the level of load which occurs between columns. It will be seen that the effectiveness of columns in the windward and leeward walls is affected by their proximity to the walls which provide the shear connection. The phenomenon is known as shear lag (see Fig 3.60).

Fig. 3.58 The braced-tube system. The action of this is similar to the framed-tube but in this case the shear load between the windward and leeward walls is carried by diagonal bracing elements.

3.6.4 Cable structures

The most efficient forms of structure are those which are stressed in pure tension.

The steel cable structure, in which the primary load carrying elements are flexible cables acting in pure tension, is an example of this type. Because these are very efficient structures they are capable of very long spans (Fig. 1.13).

The key feature of cable structures is high flexibility which allows changes in geometry to occur in response to variations in load and maintains the state of purely tensile stresses. The extent to which a geometric change can be permitted in a building is limited, however,

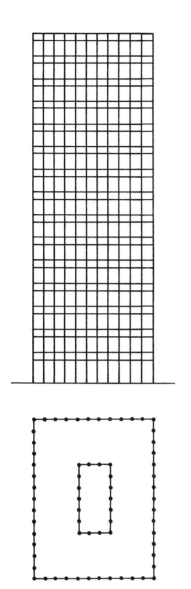

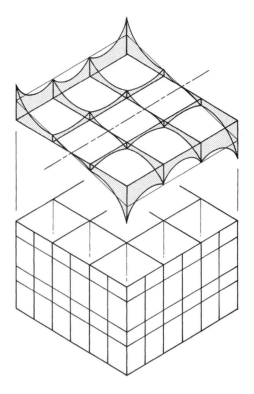

Fig. 3.60 The bundled-tube system. In this system, closely-spaced columns on the perimeter of the building act in conjunction with rows of closely-spaced columns which cross the interior. These provide additional shear connections between the windward and leeward walls and reduce the phenomenon of shear lag.

Fig. 3.59 The tube-in-tube system. In this system closely-spaced columns on the perimeter of the building act compositely with the core to resist horizontal load.

and cable structures must be designed so that drastic changes in shape, such as might occur due to a complete reversal of load under the action of high wind suction, must be prevented. A number of techniques are used to achieve this. Most are based on the use of two sets of cables, which are arranged so that the structure can resist forces from all directions

(Figs 3.61 and 3.62). Multi-cable systems have the advantage that the cables can be stressed against each other (prestressed) which further restricts the amount of movement which can occur in response to variations in load. The higher the level of prestress the more rigid does the structure become.

Although a cable structure will always take up the form-active shape[14] of the loads which act on it, this does not mean that the designer has no control over its shape, because different geometries can be produced by altering the

14 See Appendix 1 for an explanation of the term 'form-active'.

97

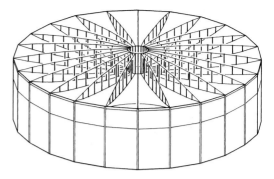

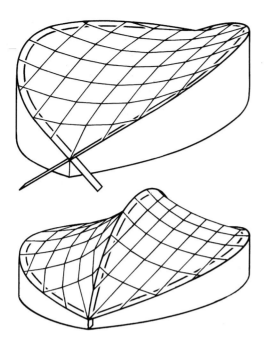

Fig. 3.62 Double-skin cable system with two sets of cables separated by short compressive elements. This type of arrangement is an alternative to single-skin systems with anticlastic shapes.

Fig. 3.61 These cable networks can resist load from different directions without undergoing gross changes in form due to their having two sets of cables which are curved in opposite directions to form an anticlastic surface. They are each supported on arch-type compressive elements. This type of cable system can also be supported on an arrangement of masts (see Fig 1.13).

support condition. Networks which are supported on masts, for example (Fig. 1.13), have a different geometry from those which are continuously supported on arches. Variations in geometry are also achieved by making use of different cable arrangements (multi-layer systems as opposed to single-layer systems, for example). The choice of geometries provided by the use of cable structures is limited, however, because all of them must be based on doubly-curved forms; all cable structures tend, therefore, to have a similar appearance.

Cable structures are highly complex, both to analyse and to construct. The difficulty in their analysis results from the high degree of statical indeterminacy which is present, due, in part, to the problem of predicting their final shape under the action of a particular load. Analysis by computer must frequently be supplemented by model analysis. Constructional detailing problems arise from the difficulty of providing suitable end anchorages for the tension members (one of the classic problems of engineering) and from the need to construct the structure in such a way that it can accommodate the small changes in geometry which will occur in response to changes in load. This normally requires the use of complicated joint components, which must be specially machined and are therefore expensive. The cladding system also must be able to adjust itself without damage to the movements which will occur and this involves the design of joints which are both weathertight and of adequate strength, and which can accommodate movement. Cladding systems tend therefore to be fairly sophisticated and to have poor durability. The consequences of all of these complexities is that cable structures are very expensive; they have been most often used for high-prestige temporary buildings such as exhibition halls.

Chapter 4

Reinforced concrete structures

4.1 Introduction

Concrete is an extremely versatile structural material. It is moderately strong in compression but weak in tension; it has good resistance to fire and good durability. Perhaps its most distinctive characteristic, however, is that it is available to the builder, on the building site, in semi-liquid form. This has two very important consequences. First, it allows concrete to be cast into a wide variety of shapes; the material itself places little restriction on form. Secondly, it makes possible the incorporation into concrete of other materials,

Fig. 4.1 Church, Vienna, Austria, 1965–76. Fritz Wotruba with Fritz G. Mayr, architects. The expressive possibilities of reinforced concrete are well illustrated here. [Photo: E. & F. McLachlan]

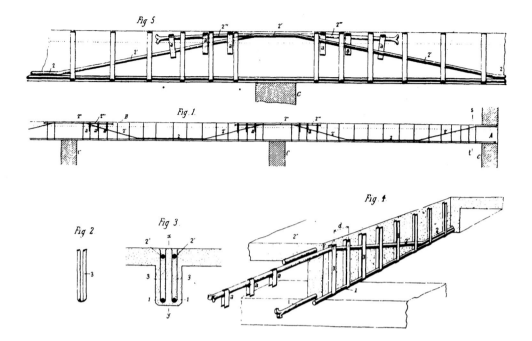

Fig. 4.2 Hennebique's system for reinforced concrete, which was patented in 1897, was one of a number which were developed in the late nineteenth century and which led to the subsequent widespread use of the material for multi-storey structures.

with which its properties can be augmented. The most important of these is steel, in the form of small-diameter reinforcing bars, and this produces the composite material reinforced concrete, which possesses tensile and flexural strength as well as compressive strength. Reinforced concrete can therefore be used to make any type of structural element.

Concrete can be either cast directly into its final location in a structure, in which case it is said to be *in situ* concrete, or used in the form of elements which are cast at some other location, usually a factory, and simply assembled on site, in which case it is referred to as precast concrete. The relative advantages of the two types are reviewed in Section 4.4. Both *in situ* and precast concrete can be produced in ordinary reinforced form or in pre-stressed form. The distinctions between these types are discussed in Section 4.3.2.

4.2 The architecture of reinforced concrete – the factors which affect the decision to select reinforced concrete as a structural material

4.2.1 The aesthetics of reinforced concrete

The opportunities which reinforced concrete offers in the matter of architectural form can be seen by examining the range of building types for which it has been used during the relatively short period in which it has been available as a structural material.

Although a type of concrete was used to good effect by the architects and engineers of Roman antiquity the modern material, reinforced concrete, dates from the nineteenth century. 'Roman' cement, the forerunner of the present-day Portland Cement, was patented in the UK by J. Aspden in 1824, but the earliest uses of reinforcement in concrete appears to have occurred more or less simultaneously in France and USA where, in the 1880s and 1890s, Francois Hennebique (Fig. 4.2) and Ernest Ransome, respectively, each developed framing systems for buildings based on this principle. Another early innovator was Robert Maillart, whose development of the two-way-spanning

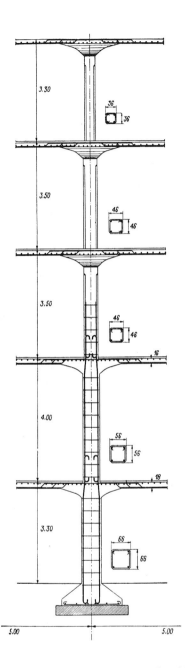

Fig. 4.3 Cross-section through five-storey flat-slab structure by Robert Maillart, *c.* 1912. Flat-slab systems, which are sophisticated two-way-spanning structures which derive much of their strength from the high degree of structural continuity which is present, were developed surprisingly early in the history of reinforced concrete construction. Because there are no downstand beams the system allows a very economical use to be made of both material and labour. Local thickening is required in the vicinity of the columns, where shear forces are high. This is accomplished by the use of 'mushroom' column heads.

flat-slab structure was particularly notable (Fig. 4.3). All of these early pioneers were concerned principally with the application of reinforced concrete framing systems to industrial buildings.

In the late nineteenth century reinforced concrete was a 'new' structural material, capable of producing durable and fire-proof skeleton frameworks and therefore buildings with open interiors free from structural walls. It arrived on the architectural scene at a time when the precursors of the Modern Movement were exploring the possibilities of creating a new architectural language which would be appropriate for the twentieth-century world. These architects were anxious to make use of the new materials which industry was producing and the most innovative of them were not slow to appreciate the potential of reinforced concrete.

Among the earliest of designers to understand the purely architectural qualities of the new material was August Perret. In the apartment block at 25 bis Rue Franklin, Paris, 1902 (Fig. 4.4), the adoption of a reinforced concrete frame structure was used to produce an open-plan interior with light non-loadbearing partition walls. Large areas of glazing were a feature of the exterior and the reinforced concrete columns of the building, although not actually exposed (a tile cladding system was used), were expressed on the facade. The later garage at 51 Rue de Ponthieu by Perret was also based on reinforced concrete and in this building the concrete framework was left entirely exposed, apart from a thin layer of paint. These buildings were very important precursors of the Modern Movement. An important aspect of their novelty was the role played by the structure in liberating the organisation of space from the tyranny of the loadbearing wall. This was exploited by Perret in his design for the Rue Franklin flats, producing a version of the Paris apartment in which a new free-flowing space could be enjoyed. The other significant aspect was the re-establishment of structure, and in particular the column, as a part of the architectural expression, something which had not happened since the eclipse of Neo-

Fig. 4.4 Apartment block, 25 bis rue Franklin, Paris, 1902–05, Auguste Perret, architect. By using a reinforced concrete frame structure Perret was able, with this building, to redefine the layout and fenestration of the traditional Paris apartment. The building also made a significant contribution to the development of a new architectural aesthetic based on the technology of the skeleton frame.

legacy from his master, but reinforced concrete was, for Le Corbusier, the ideal structural material. Its excellent structural properties placed little restriction on architectural form but above all other considerations were the newness of the material and the fact that it was a product of an industrialised process. It was therefore an appropriate medium with which to develop an architectural language for the twentieth century – the age of rationalism, industry and social idealism.

That Le Corbusier appreciated the physical properties of reinforced concrete is well demonstrated by his famous drawing of the structural core of the Domino House of 1914 (Fig. 4.5). This showed that he understood the two-way-spanning capability of the material and knew of its ability to cantilever beyond the perimeter columns. His realisation that the integral stair of this small structure could be used to brace it in the two principal directions led to the adoption of very slender supporting columns. The structure therefore caused minimal interference to the layout of the interior of the building or to the treatment of the exterior. The stair could, of course, have been positioned anywhere in the plan.

classicism by Romanticism in the early nineteenth century.

Le Corbusier, who was for a time a pupil of Perret, was another early pioneer of the architectural use of reinforced concrete. His enthusiasm for the material was, to some extent, a

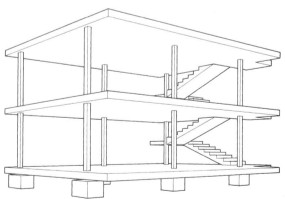

Fig. 4.5 This drawing depicts the structural carcass of Le Corbusier's Domino house of 1914 and demonstrates that he had a complete, probably intuitive, understanding of the structural potential of the material. The renowned 'five points', which were to have such a profound influence on the architecture of the twentieth century, were developed from this.

Fig. 4.6 Villa Savoye, Poissy, France, 1929. Le Corbusier, architect. The technology of the reinforced concrete framework contributed significantly to the visual language developed by Le Corbusier in the 1920s. [Photo: Andrew Gilmour]

The planning freedom which such a structural system offered the architect was summarised by Le Corbusier in his well-known 'Five points towards a new architecture'[1] of 1926 and exploited by him in the designs for the houses which he built in the 1920s, culminating in the Villa Savoye of 1929 (Fig. 4.6). Most of these buildings did not in fact have the two-way-spanning ribless slab of the Domino House but were based on beam/column frame arrangements with one-way-spanning slabs. The planning freedom

offered by the use of a concrete framework was nevertheless fully exploited.

The origins of the visual vocabulary which was evolved by Le Corbusier in the 1920s have been well documented elsewhere.[2] Among these was the idea that Modern architecture should be tectonic (in other words that the structure should influence the architecture) and that it should symbolise rationalism. It is

---

1 The renowned 'five points' were first declared in *Almanach de l'Architecture moderne*, Paris, 1926. The most significant of them were: separation of structure and skin, freedom in the plan and freedom in the treatment of the facade.

2 For example, Curtis, W. J. R. *Le Corbusier: Ideas and Forms*, Oxford, 1986.

Fig. 4.7 The Bauhaus, Dessau, 1926. Walter Gropius, principal architect. This building had a reinforced concrete frame structure which provided the same opportunities for new architectural expression as had been recognised by Le Corbusier. The glazed curtain wall which rises through the upper floors of the building represents an exploitation of one of those opportunities. [Photo: Paul H. Gleye]

perhaps as a result of the influence of the second of these that Le Corbusier restricted himself, in these early buildings, to rectilinear forms and did not avail himself of the full potential of reinforced concrete to provide almost unlimited freedom in the matter of form. The rectilinearity not only symbolised reason; it also represented a sensible and therefore rational way of using reinforced concrete to provide the structure for a building. This is an example of the 'structure

accepted' relationship between structure and architecture (see Section 2.2).

Another of the icons of early Modernism, the Bauhaus building in Dessau (1926) (Fig. 4.7) by a team of architects led by Walter Gropius, also had a reinforced concrete frame structure, the layout of which was very straightforward. In the principal block, a rectangular grid of columns supported a series of deep beams which spanned across the building and carried a one-way-spanning hollow-block floor.[3] The floor deck was cantilevered slightly beyond the perimeter columns and this allowed the steel and glass curtain wall on the exterior to run

---

3 The hollow-block floor is a form of reinforced concrete ribbed slab in which ceramic blocks act compositely with *in situ* reinforced concrete.

continuously through the three upper storeys of the building. The steel mullions which supported the curtain wall were mounted directly on the edge of the floor slab. The opaque parts of the exterior wall were constructed in masonry which was rendered to give a smooth white finish. The columns and beams of the reinforced concrete framework were left exposed throughout the interior.

The architectural vocabulary of the Bauhaus building was very similar to that which was being developed contemporaneously by Le Corbusier. Its characteristics were due, in no small measure, to the use of a reinforced concrete framework which allowed great freedom to be exercised in both the internal planning and the treatment of the exterior.

An early demonstration in Britain of the architectural qualities of reinforced concrete occurred in the Boots Warehouse at Beeston by Owen Williams (1930) (Fig. 4.8). In this case a two-way-spanning slab was adopted, supported on a square grid of columns with exaggerated mushroom heads, but visually the building was similar to the earlier examples described above. A further example was the Highpoint I apartment block by Berthold Lubetkin of the Tecton Group and Ove Arup (1933). Lubetkin and Arup had earlier demonstrated their awareness of the capabilities of reinforced concrete in the buildings which they designed for London Zoo – most notably and eloquently the Penguin Pool of 1933 (Fig. 4.9). In the Highpoint block a loadbearing-wall type of structure was adopted with one-way-spanning floor slabs. This arrangement was used to provide variations in the floor plans at different levels. Where wall-free spaces were required, the loadbearing reinforced concrete walls in the floor above were made to act as storey-height beams spanning between columns. The device is seen at its most extreme in the ground floor where the loadbearing walls were eliminated entirely and the vertical support was provided solely by a grid of columns.

The visual language developed in these highly influential early examples of twentieth-century rationalist architecture owed much to

Fig. 4.8 Boots Warehouse, Beeston, England, 1930–32, Owen Williams, architect/structural engineer. The structure here is a two-way spanning flat slab with mushroom-head columns. Its prominence and the transparency of the walls produce a similar tectonic quality to that seen in the Bauhaus (Fig. 4.7) and place the building in the forefront of contemporary architectural development. [Photo: British Architectural Library Photographs Collection]

the properties of the new structural material by which they were supported. In all of them the relationship between structure and architecture was that of 'structure accepted' (see Section 2.2), in which the visual and structural programmes were allowed to co-exist without conflict. The forms were predominantly rectilinear; this made easy the construction of the formwork in which the liquid concrete would be cast and, given the spans involved, was a sensible choice of structural form.

Rectilinearity was also an appropriate aesthetic choice given the ideas which the visual vocabulary was intended to express. In

Fig. 4.9 Penguin Pool, London Zoo, London, England, 1933. Berthold Lubetkin and Tecton, architects; Ove Arup, structural engineer. The expressive possibilities of continuous structures of reinforced concrete are most eloquently demonstrated here. [Photo: E. & F. McLachlan]

all of the buildings some parts of the structure were visible; in some buildings the structure in its entirety was visible. This option was made technically possible by the fire-resisting properties of concrete and also by its durability, and was an important aspect of the visual quality of the buildings. In addition to the fact of its exposure, the tactile nature of the exposed structure was eminently suited to the expression of the Modernist ideals of the 'honest' portrayal of the constituents of a building. Similar developments were occurring simultaneously in steel-framed buildings, as was shown in Chapter 3, but the architectural

language of reinforced concrete was subtly different from that of steel, due mainly to its different structural properties, to the fact that it could be left exposed in the finished building and to the quite different ways in which steel and reinforced concrete buildings were constructed. Concrete therefore made a distinct contribution to the developing language of early architectural Modernism.

In his buildings of the 1940s and 1950s, Le Corbusier introduced a new element into the vocabulary of reinforced concrete. This was board-marked exposed concrete ('*béton brut*').[4] All concrete does, of course, bear the marks of its formwork, but in the case of *béton brut* the

---

4 Exposed concrete which bears the marks of the formwork in which it was cast (literally 'rough', 'raw', 'unfashioned', 'unadulterated').

Fig. 4.10 Monastery of La Tourette, Eveux, France, 1955. Le Corbusier, architect. Exposed structural concrete, which bears the marks of the rough formwork against which it was cast, was a notable aspect of the architectural vocabulary employed by Le Corbusier in this period. It was compatible with his architectural thinking at that time but was also ideally suited to the requirements of this particular client and project. [Photo: Andrew Gilmour]

formwork was very crudely constructed and produced textures which were rough and uneven. The device was first used by Le Corbusier in buildings such as the 'Petite Maison de Weekend' of 1935 but the most notable of his buildings in which it was employed were the 'Unité d'Habitation', Marseilles, of 1947–52 and the monastery of La Tourette, near Lyons, of 1957–60 (Fig. 4.10). In these later buildings the crudity of the construction was to some extent caused by the

inexperience of the building contractors but there is no doubt that Le Corbusier himself found that the suggestions of 'savageness' and 'the primitive' which it produced were compatible with his architectural thinking at that time.

The resulting style was to be much imitated, a prominent British example being the National Theatre of Great Britain (1967–76), London (Fig. 4.11) by Denys Lasdun, and was entirely compatible with the fashion for 'New Brutalism' which had emerged in Britain in the

Fig. 4.11 National Theatre of Great Britain, London, 1967–73, Denys Lasdun, architect. This building is a very prominent example of the type of Corbusier-inspired architecture which was fashionable in the 1950s and 1960s in which the expressive possibilities of roughly textured exposed reinforced concrete were explored. (Photo: Densy Lasdun Peter Softley and Associates).

1950s. Another British example was the Economist Building, again in London (1964), by Alison and Peter Smithson. Frequently, the textures involved were deliberately contrived to be extremely rough and thus was a new and savage element introduced into the otherwise smooth and refined vocabulary of architectural Modernism.

The architectural language developed by Le Corbusier, Gropius, Lubetkin, Williams and others in the 1920s and 1930s, and extended in the 1940s and 1950s by Le Corbusier, Alison and Peter Smithson and others, was destined to be imitated in almost every city in the world. In virtually all of the buildings involved, whether the seminal buildings in which the language was evolved or the many imitations, the relationship between the structure and the architecture fell into the category of 'structure accepted'. The properties and requirements of the reinforced concrete structure were, in other words, given an important role in influencing the visual language of the architecture.

Not all of the great reinforced concrete buildings of the twentieth century fall into the category described above, however. Le Corbusier himself did occasionally make full and exaggerated use of the expressive possibilities of reinforced concrete; perhaps the most famous example of this occurred at the chapel at Ronchamp (Fig. 4.12). Here the relationship between the structure and the architecture was one of 'structure ignored'.[5] Structural considerations played little part in the evolution of the form of the building. The structural system which was adopted was nevertheless relatively straightforward, largely due to the excellent structural properties of reinforced concrete. The distinctively shaped roof is nothing more than a beam/column framework with a one-way-spanning slab. In this case the mouldability of the concrete and the structural continuity it offered were used to produce a slab with double curvature to form the remarkable canopy. The beams which support this were

5　See Section 2.2.

concealed by the upward sweep of the roof and the columns were buried within the self-supporting masonry walls.

It should be noted, however, that the relative ease with which this complex form was achieved, using a very basic type of structure, was due to the small span of around 20 m. Had the building been significantly larger, a more efficient type of structure would have been required ('semi-form-active' or 'form-active'). In that case it might, for structural reasons, have been necessary to adopt a slightly different shape of roof and Le Corbusier would have found himself in the position of having to accept restrictions on his total freedom to invent the form of the building.

Another famous building which illustrates well the excellent structural properties of reinforced concrete is the 'Falling Water' house by Frank Lloyd Wright (1936) (Fig. 4.13). As with the chapel at Ronchamp this is an example of 'structure ignored'. The oversailing and apparently free-floating horizontal

Fig. 4.12 Notre-Dame-du-Haut, Ronchamp, France, 1954. Le Corbusier, architect. This form, with its elaborate roof canopy which does not actually touch the tops of the walls (there is a gap between roof and walls which is used to good effect to introduce light into the interior) could only have been constructed in reinforced concrete. It illustrates very well the freedom to invent form which the material confers on the architect. [Photo: P. Macdonald]

platforms which are one of the principal features of this architecture, and which are essential for conveying the ideas of free-flowing space, the interpenetration of solid and void and the blurring of the boundary between exterior and interior, were not ideal from a structural point of view. Their feasibility depended on the use of two-way-spanning reinforced concrete slabs supported on a network of columns. As at the Villa Savoye, the decks are not solid slabs but networks of beams and cantilevers supporting a thin floor slab. Some of the cantilevered beams are of massive proportions and, as at Ronchamp, the

109

Fig. 4.13 Falling Water, the Bear Run, Pennsylvania, 1936. Frank Lloyd Wright, architect. The cantilevered balconies of this building required the use of a high-strength structural material which would allow a high degree of structural continuity to be developed. Reinforced concrete was the ideal material in which to realise this form. [Photo: Andrew Gilmour]

size of the building is close to the limits of what is feasible with the method of construction employed.

It is interesting to note that, unlike the case of the rationalist buildings of Le Corbusier, Gropius, Lubetkin and others, in which the logic of the structure was accepted and allowed to influence the architectural language, neither of these buildings in which structure was virtually ignored when the form was determined was destined to be imitated to any significant extent.

In the building boom which followed the war of 1939–45 many buildings which conformed to the mainstream of architectural Modernism, as it had been developed in the 1920s and 1930s, were erected with reinforced concrete structures. The concrete was often left exposed and expressed on the exterior, especially following the advent of the fashion for 'Brutalism'. Much of this architecture was of indifferent quality, however, and as a consequence, the term 'concrete jungle' was added to the language of architectural criticism, albeit not by mainstream architectural critics who were, and remain, largely apologists for this type of architecture. Even in the best examples, the full potential of reinforced concrete as a material which allows the creation of almost every type of form was not realised. The preferred shapes were almost invariably rectilinear.

The language of reinforced concrete architecture continued to develop, however. The distinctive buildings designed by James Stirling in the late 1950s and 1960s, for example the Leicester Engineering Faculty Building (1959–63) (Fig. 4.14), The History Faculty Building at Cambridge University (1964–67) and the Florey Building at Oxford University (1966–71) (Figs 4.15 and 4.16), all in the UK, represent a more adventurous exploration of the expressive potential of reinforced concrete for multi-storey institutional buildings than was seen in the period of early Modernism.

The form of the Florey Building (Figs 4.15 and 4.16) is distinctive in both plan and elevation. The structure is a reinforced concrete framework and consists of a series of identically shaped main frames each with canted columns and horizontal beams carrying the one-way-spanning floor decks. Several design features, namely the crescent-shaped plan composed of a series of rectangles and triangles whose boundaries are defined by the main frames (Fig. 2.9), the distinctive cross-section and the ingeniously moulded stairs which penetrate the sloping exterior skin of the building (Fig. 4.16), create many junctions between the elements which would have been very difficult to fashion in any material other than *in situ* reinforced concrete. The structural continuity which the material allows was also an essential requirement for the complex geometry of this building. The structural system is working here well within its capabilities and the fact that it has been left exposed to form a prominent part of the architectural language, places the building

Fig. 4.14 Engineering Building, Leicester University, England, 1965–67. James Stirling, architect, F. J. Samuely & Partners, structural engineers. This building has a reinforced concrete framework structure and demonstrates two of the principal advantages of the material. Its high durability has allowed the structure to be exposed on both the exterior and interior, and thus to contribute to the architectural language being used, and its mouldability has been employed to produce a relatively complex building form (Photo: P. Macdonald).

firmly within the rationalist tradition of Gropius and early Le Corbusier.

A more recent building in which the structural properties of reinforced concrete have been fully exploited is the Willis, Faber and Dumas Building at Ipswich by Foster

Fig. 4.15 Accommodation plan and cross-section of the Florey Building, Oxford, England, 1967–71. James Stirling, architect, F.J.Samuely & Partners, structural engineers. The distinctive plan and cross-section of this building were easily realised with a structure of *in situ* reinforced concrete.

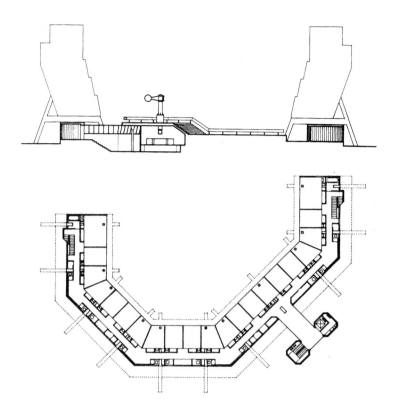

Associates (1975) (Fig. 4.17). This building is of the same basic constructional type as the early buildings of Gropius and Le Corbusier. It consists of a reinforced concrete structural armature giving a plan which is free from structural walls. Two-way-spanning coffered slabs, supported on a square column grid, define the three principal floors. These are readily visible from the outside through the transparent skin which is entirely of glass and which is mounted directly on the thin edges of the floor slabs. Both the distinctive curvilinear plan of this building and the cantilevering of the floor slabs beyond the perimeter columns are expressive of the unique properties of structural concrete and recall Le Corbusier's drawing of the structural armature of the Domino house (Fig. 4.5).

The architectural language of reinforced concrete was greatly extended in the 1980s,

particularly by the work of the New York Five.[6] The Museum of Decorative Arts, Frankfurt on Main (1979–85) by Richard Meier (Fig. 4.18), will serve as an example. In this building Meier accepted the vocabulary of early Modernism but not the grammar or syntax, in the sense that he did not subject himself to a dogmatic belief in a quest for a universally applicable architecture, as Gropius and Le Corbusier had done. Instead he allowed the requirements of the building's situation and of the very complex brief to affect the development of the form.

The plan of the building was based on two rectangular grids set at an angle of 3.5° to each other (Fig. 4.19). Its essential features are a grouping of distinctively shaped masses around courtyards, giving a juxtaposition of solid and void which is seen also in the alternation of solid and transparent sections in the external skin. The structure is basically a post-and-beam framework but the geometries of the plan and elevations are complex. These, and the bold treatment of small-scale features such as doorways and balconies, are all readily accom-

---

6 Richard Meier, Peter Eisenmann, Michael Graves, Charles Gwathney, John Hejduk.

Fig. 4.16 Florey Building, Oxford, England, 1971. James Stirling, architect, F.J.Samuely & Partners, structural engineers. The stairs which protrude from the inwardly canted rear wall of the Florey Building provide a good illustration of the type of complexity which the mouldability and structural continuity of a reinforced concrete structure allow (Photo: P. Macdonald).

Fig. 4.17 Willis, Faber and Dumas office, Ipswich, England 1974. Foster Associates, architects, Anthony Hunt, structural engineer. The curvilinear plan, cantilevered floor slabs and transparent walls of this building give architectural expression to the structural properties of reinforced concrete (Photo: Alastair Hunter).

Fig. 4.18 Museum of Decorative Arts, Frankfurt on Main, Germany, 1979–85. Richard Meier, architect. The juxtaposition of solid and void and of rectilinear and curvilinear elements is typical of the work of Richard Meier. *In situ* reinforced concrete is an ideal structural material for this type of architecture [Photo: E. & F. McLachlan]

modated by the continuity and mouldability of the reinforced concrete structure. In both its overall form and in its detailing the building produces constant reminders of the sculptural possibilities of the structural material and is therefore another example of the contribution which reinforced concrete has made to the developing language of mainstream modern architecture. Meier produced here a building which functioned well for its intended purpose, which was a positive addition to the city of Frankfurt, and which extended the architectural language of reinforced concrete.

Another example of the use of reinforced concrete to produce a highly sculptural form is the building for the Vitra Design Museum (1988–89) (Fig. 4.20) by Frank Gehry. As with earlier examples of this genre, such as Le Corbusier's chapel at Ronchamp, this is a relatively small building and the ability of reinforced concrete to allow the architect almost unlimited freedom in the matter of form, provided that a building is not too large, is well demonstrated.

Buildings such as Le Corbusier's chapel at Ronchamp and Gehry's Vitra Design Museum might be described as belonging to the 'irrational school' of curvilinear concrete architecture. There has also been a 'rationalist school' which has produced structures which are entirely justifiable in a purely technical sense as well as being striking visually. Such structures have normally been the designs of structural engineers (Robert Maillart, Pier Luigi Nervi, Felix Candela, Eduardo Torroja, Santiago Calatrava) rather than of architects.

This group of buildings is characterised by forms which allow a very high ratio of strength to weight to be achieved, principally by

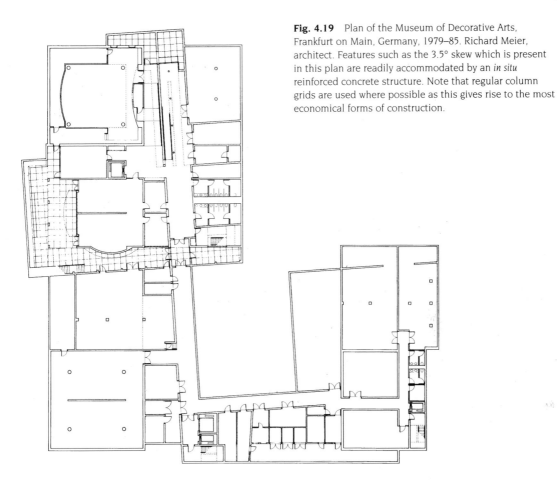

Fig. 4.19 Plan of the Museum of Decorative Arts, Frankfurt on Main, Germany, 1979–85. Richard Meier, architect. Features such as the 3.5° skew which is present in this plan are readily accommodated by an *in situ* reinforced concrete structure. Note that regular column grids are used where possible as this gives rise to the most economical forms of construction.

conforming to 'form-active'[7] shapes. Many, for example, are based on the parabola, which is the 'form-active' shape for uniformly distributed gravitational load. They are also shapes which can be easily described mathematically, and this makes both their design and their construction much simpler than those of equivalent irregular forms such as the chapel at Ronchamp or the Vitra Design Museum. The design is simpler because the analysis of the structure, in which internal forces and stresses are calculated, is straightforward if the form can be described mathematically by equations based on a Cartesian co-ordinate system. The construction is straightforward when the form

can be described mathematically because this greatly eases the problem of setting out and building the formwork on which the concrete is cast. Thus, the buildings which have been designed by the architect/engineers fall into the category of 'true structural high tech' (see Section 2.2). They perform well when judged by purely technical criteria concerned with structural efficiency and 'buildability' in present day industrialised societies because their shapes are both 'form-active' and part of the world of precise mathematical description rather than of personal preference.

To sum up, the buildings which have been described above serve to illustrate both the contribution which reinforced concrete has made to the development of twentieth-century architecture and the range of architectural forms made possible by the material.

7 See Appendix 1.

Fig. 4.20 Vitra Design Museum, Basel, Switzerland, 1987–89. Frank Gehry, architect. This highly sculptural form could only have been realised in reinforced concrete [Photo: E. and F. Mclachlan].

Reinforced concrete has been used principally in the form of skeleton frameworks where it has allowed architects to exploit the opportunities articulated in Le Corbusier's 'Five points towards a new architecture' of 1926, in particular, the '... free designing of the ground plan ...' and the '... free designing of the facade ...'. Examples of this type of building can be found in every decade of the Modern period, from Le Corbusier and Gropius to Foster and Meier. In most cases the relationship between structure and architecture has been one of 'structure accepted' (see Section 2.2) in which the reinforced concrete armature of the building, though playing a significant role in determining the form and general arrangement, was not emphasised in the aesthetic treatment, which was primarily, if not exclusively, orthogonal. Some notable exceptions to the rectilinear geometry which characterised the mainstream Modern Movement have also been shown. In these the mouldability of concrete was exploited to produce distinctive irregular or curvilinear structural shapes. This type of building, which has become more common in recent decades as a result of changing architectural fashion, illustrates better the full potential of reinforced concrete as a structural material.

4.2.2 The technical performance of concrete as a structural material

4.2.2.1 Introduction
The technical advantages and disadvantages of concrete in relation to those of alternative materials are reviewed in this section. These

are obviously an important consideration in relation to the selection of concrete as the structural material for a building.

4.2.2.2 Advantages

Strength

Of the four principal structural materials reinforced concrete is one of the strongest. It performs well in skeleton-frame-type structures and is therefore best used in situations in which the properties of a frame are required (planning situations in which the restrictions of loadbearing walls cannot be accepted). It is particularly suitable for frameworks on which the level of imposed loading is high.

Reinforced concrete has tensile, compressive and flexural strength and can be used to make all types of structural element. It is capable of resisting the internal forces which result from every possible combination of applied load and structural geometry and is therefore capable of producing elements with any geometry. The material itself places no restriction on structural form.

Mouldability

The fact that reinforced concrete is available in semi-liquid form means that it can be cast into an almost infinite variety of shapes. This property, together with its strength characteristics, means that virtually any form can be created relatively easily in reinforced concrete. The moulding process also allows structural continuity between elements to be achieved relatively easily and the resulting statical indeterminacy is another factor which eases the production of complicated forms. In particular, it allows the adoption of irregular patterns of vertical support for floor and roof structures, the cantilevering of floor and roof structures beyond perimeter columns and the omission of areas of floor to create voids running through more than one storey in the interiors of buildings. It also allows reinforced concrete structures to be self-bracing.

Durability

Reinforced concrete is a durable material which can be left exposed in relatively hostile environments. A wide variety of surface textures can be achieved, depending on the type of mould treatment which is specified. Finishing materials can therefore be eliminated where concrete structures are used.

Fire resistance

Reinforced concrete performs well in fire; it is incombustible and it retains its structural properties when exposed to high temperatures.

Cost

Reinforced concrete is relatively cheap and, when used for frame structures, will usually be cheaper than steel. It is, however, normally more expensive than masonry for loadbearing-wall structures.

4.2.2.3 Disadvantages

Weight

Reinforced concrete structures are heavy. The material has a relatively low ratio of strength to weight and a reinforced concrete frame is normally significantly heavier than an equivalent steel frame. It is because of the high self-weight that reinforced concrete performs best in situations in which relatively high imposed loads are involved – for example, for the floor structures of multi-storey commercial or industrial buildings. It is rarely used where imposed loads are small, such as in single-storey buildings or the roof structures of multi-storey buildings, except in forms in which the level of structural efficiency is high, such as form-active shell roofs.

The relatively high self-weight of reinforced concrete structures does have some advantages, however. It gives the building a high thermal mass which eases the problems associated with environmental control. It also means that concrete walls and floors are capable of acting as effective acoustic barriers.

Construction

The construction of a reinforced concrete structure is complicated and involves the erection of formwork, the precise arrangement of intricate patterns of reinforcement and the careful placing and compacting of the concrete itself. The construction process for a reinforced

117

concrete structure therefore tends to be both more time consuming and more costly than that of an equivalent steel structure and these factors can mitigate against its use in a particular situation. Another disadvantage is the requirement for sufficient space for storage of formwork and for the assembly of reinforcement cages. This can be problematic if the building site is very tight and congested.

Strength
Although, as stated above, reinforced concrete is one of the strongest of the four primary structural materials it is nevertheless weaker than steel, with the result that elements in reinforced concrete structures tend to be bulkier than steel equivalents. The relative weakness also places restrictions on the spans for which reinforced concrete is suitable. A practical maximum span for floor structures is around 20 m and spans greater than 15 m are in fact rare.

4.2.3 The selection of reinforced concrete
A structural material is selected on the basis of the aesthetic opportunities which it allows and the technical performance which it provides. Many of the former are, of course, consequences of the latter.

So far as the overall form of a building is concerned, most of the possibilities which reinforced concrete allows result from both its relatively high strength and the high level of structural continuity which it makes possible. The high strength makes it suitable for skeleton-frame-type structures in which the internal forces are relatively high. It is most suitable, however, for structures in which fairly large imposed loads are carried over relatively short spans, that is spans in the range 6 m to 15 m, and is therefore employed mainly for multi-storey buildings where floor loads have to be carried.

The high level of structural continuity which the use of reinforced concrete allows gives the designer considerably more freedom to manipulate the overall form of a framework than is possible with steel (the principal alternative). The two-way-spanning capability of the

floor structure is especially significant and allows both the adoption of an irregular pattern of vertical support and the omission of sections of floor to create volumes which run through more than one storey. It also makes possible the cantilevering of floors beyond perimeter columns and the simple creation of ramps or stepped changes in the levels of floors. The structural continuity of reinforced concrete also facilitates the creation of curvilinear plan-forms, either in the complete building or in the form of internal breaks in the floor structure, and the adoption of complex, non-rectilinear column grids. Reinforced concrete is also ideally suited to the creation of form-active or semi-form-active types of structure such as shells, vaults, domes and arches. Reinforced concrete therefore offers the architect very great freedom in the matter of form.

The durability and fire resistance of concrete are two very significant properties if the expression of structure is an aspect of the architectural programme. The range of surface textures with which concrete can be produced is an added benefit where this is done. Even where the structure is not exposed, these qualities will normally simplify both the initial construction of a building and its subsequent maintenance by removing the necessity for fireproofing and corrosion protection schemes.

4.3 A brief introduction to concrete technology

4.3.1 Introduction
Reinforced concrete is a composite material whose constituents are concrete, which forms the main bulk of the material, and steel, in the form of reinforcing bars. Concrete itself is also a composite material being composed of cement and aggregate (fragments of stone). The properties of concrete depend on those of its constituents and on the proportions in which these are mixed. Concrete can actually be manufactured on the building site although in modern practice it is normal for even *in situ* concrete to be mixed in a separate factory and

delivered to the site in liquid, ready-mixed form. The mix proportions can, however, be specified by the building's designer and concrete is therefore one of the few materials whose properties can be controlled directly by the designers of buildings. The properties of reinforced concrete depend on those of the constituent concrete and of the reinforcement and on the location of the reinforcement in the structural elements.

Reinforced concrete is therefore a complex material which places certain demands on the designer who wishes to exploit its potential fully. The successful use of such a material must be based on a knowledge of its basic properties and its behaviour in response to load. These aspects of concrete technology are not considered here in detail, although the principal issues are briefly described.

4.3.2 The terminology of concrete

The constituents of concrete are cement paste, which acts as a binding agent, and aggregate, which is an inert filling material. The aggregate usually consists simply of small pieces of natural stone and it acts both as a bulking agent and to impart dimensional stability to the concrete. Concrete is made by mixing together appropriate quantities of cement and aggregate in the dry state and adding sufficient water to hydrate the cement. After the water is added, a chemical reaction occurs which causes the concrete to become solid within a few hours in what is called the 'initial set'. A considerable period of time is required before it develops its full strength, however, and this latter process is called the 'final setting' or 'hardening' of the concrete. The time required for final setting varies, depending mainly on the type of cement which is used, but a typical concrete will have developed about 80% of its full strength within three months of the initial set.

The properties of concrete which are of principal interest to the building designer are its liquidity when it is in the 'fresh' state and its strength when in the hardened state. The liquidity of fresh concrete is referred to as its 'workability' and this property affects the ease

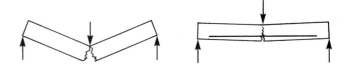

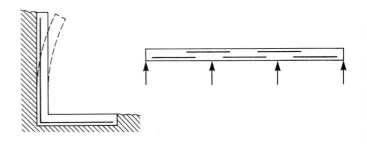

Fig. 4.21 The principal reinforcement in reinforced concrete is placed in the locations where tension occurs due to the bending effect of the load. Secondary reinforcement (not shown) is placed throughout the structural elements to resist the tensile stress which occurs due to the shrinkage associated with the curing process and with thermal movement.

with which it can be 'placed' and 'compacted' to form a dense solid. This is an especially important consideration where elements of complex geometry are being constructed or where the concrete must be compacted around complicated patterns of reinforcement. High workability is required in these situations and this can adversely affect the final strength of the concrete (see Section 4.3.3.3).

Plain concrete in the hardened state is a material which has moderate compressive strength (typically between 20 N/mm^2 and 60 N/mm^2 depending on the mix proportions) but very low tensile strength (usually about one tenth of the compressive strength). When steel bars are incorporated into concrete the resulting composite material is called reinforced concrete. This has greatly improved tensile and bending strength, the precise strength properties being dependent on the amount of reinforcement which is used and its location in element cross-sections. In elements

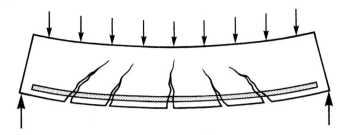

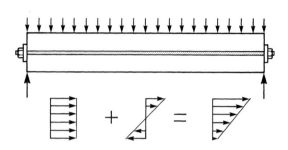

Fig. 4.22 Cracking of the concrete, shown exaggerated here, is inevitable in the parts of reinforced concrete elements in which tensile stress occurs. The width of the cracks which form must be controlled to prevent the ingress of water, which would cause corrosion of the reinforcement. This places a limitation on the amount of strain, and therefore stress, which can be tolerated in the reinforcement. Allowable stresses in reinforcement are normally lower than the strength of steel would in other circumstances allow.

Fig. 4.23 The principle of pre-stressed concrete. In the top diagram the steel bar is tensioned by the bolts at the ends of the beam and introduces a compressive 'pre-stress' into the concrete which is uniformly distributed within each cross-section. This neutralises the tensile bending stress which occurs in the lower half of the beam when the load is applied and thus eliminates tensile cracking of the concrete. A quite different relationship between steel and concrete exists here than occurs in plain reinforced concrete.

which carry bending-type load the reinforcement is concentrated in locations where stretching, and therefore, tensile stress, occurs (Fig. 4.21).

The tensile internal forces which are present in reinforced concrete as a result of bending action, and which are resisted by the reinforcement, cause tensile strain[8] to occur. This produces cracking in the concrete surrounding the reinforcement which affects the appearance and durability of the elements (Fig. 4.22). The extent of the cracking must be limited by suitable design and is usually controlled by restricting the amount of stress which is permitted in the reinforcement. This limits the amount of stretching which can occur and therefore the width of any cracks which form but it has the disadvantage of requiring that the steel be greatly understressed and therefore inefficiently used.

Pre-stressing of concrete allows the full potential strength of the steel reinforcement to be realised. In pre-stressed concrete, crack widths are limited by deliberately introducing an axial compressive load into the concrete so

as to limit the amount of stretching which can develop when a bending-type load is applied. The pre-stressing is carried out by stretching the reinforcement during the construction of an element and then connecting it to the concrete in such a way that it imparts a permanent compressive stress into the concrete (Fig. 4.23). If a load occurs on the structure which tends to cause tension to develop, the compressive pre-stress must first be overcome before any stretching can occur. The pre-stressing technique thus uses the concrete–steel combination in such a way that the full tensile strength of the steel is utilised without the penalty of excessive stretching and therefore of cracking of the concrete being incurred.

Pre-stressed concrete is either 'pre-tensioned' or 'post-tensioned'. In pre-tensioning the reinforcing bars, which are usually

---

8 For definitions of stress and strain see Macdonald, *Structure and Architecture*, Appendix 2.

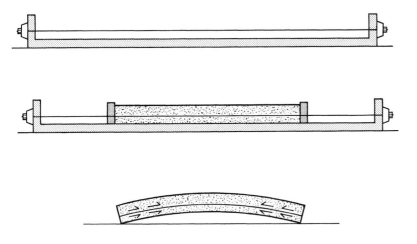

Fig. 4.24 Pre-tensioned pre-stressed concrete. In pre-tensioned pre-stressed concrete the pre-stressing wires are stretched in a pre-stressing bed prior to the placing of the concrete. The wires are released from the pre-stressing bed once the concrete has hardened and the pre-stress is transferred to the concrete by frictional (bond) stresses acting between the concrete and the surfaces of the wires.

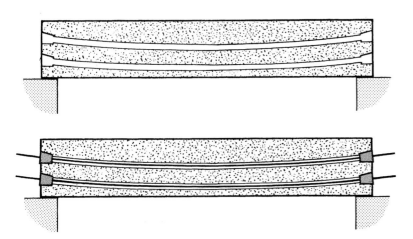

Fig. 4.25 Post-tensioned pre-stressed concrete. Where concrete is post-tensioned, ducts for pre-stressing tendons are cast into the elements concerned. The tendons are fed into the ducts once the concrete has hardened and then stretched by external jacking devices. The resulting pre-stress is then introduced to the concrete via special anchoring devices at the ends of the element.

small-diameter wires of high-tensile steel, are stretched to a predetermined tensile stress in a pre-stressing bed (Fig. 4.24). The fresh concrete is then placed around these and allowed to set and harden. The pre-tension is then released and the friction (bond) between the concrete and the surface of the reinforcement transfers a permanent compressive load into the concrete. Pre-tensioned elements are normally precast and the technique is commonly used for producing a range of proprietary components such as lintels, beams of various kinds, and floor slabs. These are used in a number of different types of structure. The lintels are normally used in conjunction with masonry.

Pre-stressed floor slab units are used in many types of structure including steel and reinforced concrete frames and masonry loadbearing-wall structures.

In post-tensioned pre-stressed concrete, which is usually cast *in situ*, the concrete is placed into moulds which contain narrow tubes (Fig. 4.25). After the concrete has set and hardened the reinforcement is placed in these tubes and tensioned against plates or jacks which are permanently fixed to the ends of the elements. This has the effect of introducing a permanent compressive pre-stress.

Pre-stressing allows much higher stresses to be used in the reinforcement in concrete and

121

permits much lighter and more slender elements to be achieved than is possible with ordinary reinforced concrete. The quality of both the concrete itself and the reinforcement must, however, be higher than in ordinary reinforced concrete and the components are generally more sophisticated. Pre-stressed concrete is usually more expensive to construct than equivalent reinforced concrete and, with the exception of the range of pre-tensioned proprietary components which has been mentioned above, the technique tends to be used for architectural structures only where a special requirement exists, such as the need to achieve a very long span.

4.3.3 The constituents of reinforced concrete

4.3.3.1 Cement
Cement is the binding agent in concrete which possesses the cohesive strength to hold the aggregate and reinforcement together into a solid, composite material. A considerable number of types of cement are used in building; most of these are varieties of Portland cement.[9] All of the cements which are used for structural concrete are dependent on water for the development of strength; when water is added to dry cement a fairly complex series of chemical reactions takes place in a process which is called hydration and which causes the resulting paste to stiffen. The subsequent development of strength takes place in two stages: initial setting occurs fairly quickly, usually within a few hours; a much longer period, of up to a year, is required for the development of full strength, although a reasonable proportion of the final strength will normally have been developed within seven days. The hardening of the cement is simply a continuation of the hydration process which produced the initial set and the cement must be kept wet if this is to proceed satisfactorily to compensate for water which may be lost due to evaporation.

One of the principal factors which affects the strength of hardened cement is the water–cement ratio, which is the ratio of the weight of water to the weight of cement in the wet mix. Only a small quantity of water (around 25% by weight) is required to bring about sufficient hydration of the cement to cause the initial set; if more is present the extra water remains as a separate phase in the cement paste and becomes incorporated as a separate phase into the cement matrix when setting occurs. The extra water subsequently evaporates to leave voids in the hardened cement and this reduces its strength. It is therefore necessary to ensure that no more water than the minimum required to produce initial setting is present in a mix of concrete.[10] Figure 4.26 shows the relationship between water–cement ratio and final compressive strength for a typical concrete.

The use of a low water–cement ratio has the effect of making the concrete very stiff and difficult to compact. This is a factor which has to be considered by structural designers when they specify the water–cement ratio which is to be used for a concrete, because the entrainment of air into the concrete, due to poor compaction, will reduce its strength. The minimum water–cement ratio which is practicable in a particular case depends on individual circumstances, such as the complexity of the element in which the concrete is to be placed and the type of equipment which will be available for use in the compaction process. The practical minimum for water–cement ratio is usually in the range 0.4 to 0.5 (40–50% by weight).

Two phenomena which are associated with the setting and hardening of cement and which affect the specification of concrete are shrinkage and heat gain. The cement matrix which is formed during the initial setting process

---

9 So called because of its supposed resemblance to Portland stone.

10 More water must in fact normally be present to produce sufficient workability (liquidity) to allow the cement to be compacted effectively. The practical minimum amount of water is around 40% to 50%, depending on the system being used for compaction.

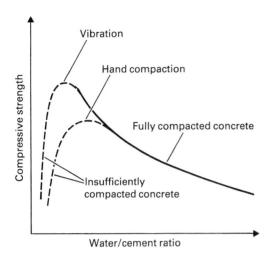

Fig. 4.26 The relationship between the ultimate strength of concrete and the water/cement ratio. This graph demonstrates that the more fluid the wet concrete is the lower will be the ultimate strength. This is because relatively little water is required to cause the initial set of concrete. If more is present it forms separate pockets of pure water in the setting concrete which ultimately evaporate to leave small voids which weaken the concrete.

shrinks subsequently as the hardening process proceeds. Shrinkage, by itself, is not necessarily harmful to a concrete component, especially if the element is not restrained during the hardening process (a small precast concrete lintel, for example). Where an element is restrained, however, such as an *in situ* reinforced concrete beam or slab which is part of a large frame, then the effect of shrinkage is to generate tensile stress in the element which may cause it to crack. Shrinkage is therefore a phenomenon which must be considered during the design of a concrete structure. Normally it is controlled by the incorporation into the structure of suitable reinforcement; this simply carries the stress which results from the shrinkage. Occasionally, movement joints are used to control shrinkage cracking. A typical example of this is found in lightly reinforced non-structural ground floor slabs.

Heat gain is a phenomenon which occurs because the chemical processes which are involved in the hydration of cement are exothermic. The increase in temperature which results from this tends to accelerate the process of hydration and this can produce cracking of elements because it also causes high rates of shrinkage. The heat of hydration rarely causes a problem in architectural structures, however, because the elements are normally slender enough to allow sufficient cooling to occur to maintain temperatures at an acceptable level. Where an element is of large bulk the use of special 'low-heat' cements is sometimes required.

4.3.3.2 Aggregate

Aggregate, being cheaper than cement, is used as a bulking agent in concrete and, typically, will account for 75% to 80% of its volume. It also serves to control shrinkage and to improve dimensional stability. It normally consists of small pieces of stone, of various sizes in the form of either naturally occurring sand or gravel, or crushed rock fragments; other materials, such as crushed brick, blast furnace slag or recycled building materials, are sometimes used. Aggregate must be durable, of reasonable strength, chemically and physically stable, and free of constituents which react unfavourably with cement.

The proportions which are present of the differently sized particles which occur in an aggregate are referred to as its grading and if the aggregate is to be effective as a bulking agent the grading must be such that a particular distribution of particle sizes occurs. The smaller particles should ideally be of such a size and quantity as to fully occupy the voids between the larger particles and leave a minimum volume to be filled by cement (Fig. 4.27). The grading of aggregate also affects the workability of the concrete, which is better when a higher proportion of fine particles is present. This is a factor which indirectly affects strength because a well-graded aggregate allows a required workability to be achieved with a lower water–cement ratio than a poorly graded aggregate.

If a naturally occurring aggregate is used for concrete, the designer has no control over the grading but must ensure that the range and

123

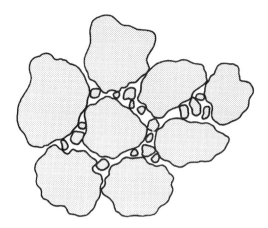 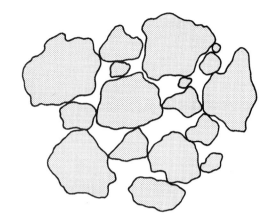

Fig. 4.27 If aggregate is to form an effective bulking agent the relative sizes of the particles present must be well matched. This has occurred in the sample on the left here, where only small voids are left to be filled with cement, but not in that on the right.

distribution of particle sizes is satisfactory. This is determined by carrying out a sieve analysis, which involves passing a sample of the aggregate through a series of progressively smaller sieves and noting the fraction which is retained on each (Fig. 4.28). The percentage of the whole sample which passes each sieve is then plotted to give a grading curve (Fig. 4.29). The relevant British Standard (BS 882) sub-divides aggregates into coarse aggregate, in

which particles larger than 5 mm in diameter predominate, and fine aggregate or sand, in which particles smaller than 5 mm in diameter predominate. The grading limits for each type which are considered to be suitable for making concrete are specified in the Standard by defining zones within which the grading curves of the fine and coarse aggregates must fall.

Where a naturally occurring aggregate with a suitable grading is not available, an aggregate with a specific grading can be produced artificially by mixing together appropriate quantities of screened, single-sized batches of crushed rock fragments.

Aggregate particles are classified into three categories according to their shape and are said to be either rounded, irregular or angular.

Fig. 4.28 The grading of a naturally occurring aggregate is determined by a sieve analysis. In the nest of sieves shown the mesh size diminishes down the stack. To carry out the analysis the sample is introduced at the top and the weight of particles which are retained at each level determined. This allows grading curves to be plotted.

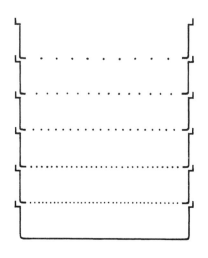 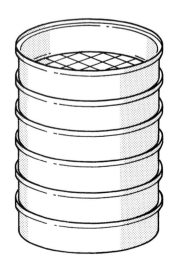

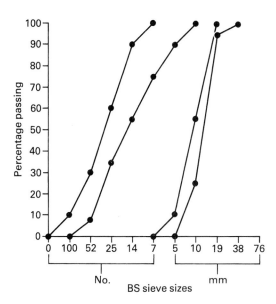

Fig. 4.29 Aggregate grading curves. Following sieve analysis a sample of aggregate will produce a grading curve similar to those depicted. Its suitability as a bulking agent in combination with other aggregates can then be judged.

Naturally occurring aggregates tend to be rounded while crushed rock types are predominantly angular. The significance of this factor is that it affects the workability of the concrete: aggregates with rounded particles produce a more workable mix for a given water–cement ratio than do those with angular particles. It is therefore possible to achieve a particular workability with a lower water–cement ratio if the aggregate is rounded than if it is angular, and this in turn produces a stronger concrete. Aggregate shape, like grading, therefore has an indirect effect on concrete strength and, if other things are equal, naturally occurring aggregates will produce stronger concrete than artificially produced aggregates.

The grading of an aggregate can affect the cost of concrete because it affects the quantity of cement which must be provided to produce a given volume of concrete. If a large number of fine particles are present, the total surface area of the aggregate is high and the quantity of cement which is required to coat all of the

particles is larger than if the same bulk is provided with a smaller number of larger particles.

A final consideration in relation to the aggregate which is used in concrete is the maximum particle size of the coarse fraction. This must usually be restricted, depending on the size of the elements in which the concrete will be cast and on the complexity of the reinforcement pattern. The maximum size of aggregate is normally no greater than 25% of the minimum thickness of elements and at least 5 mm less than the clear distance between reinforcing bars.

4.3.3.3 Specification of concrete
The specification of concrete has two aspects, which are the selection of the constituents, that is of the type of cement and type of aggregate, and the determination of the mix proportions. The latter process is called the 'mix design'.

So far as the selection of materials is concerned the type of cement is determined by the requirements in respect of rate of hardening, heat gain and resistance to chemical attack; all Portland cements achieve approximately the same final strength for a given set of mix proportions. It is usual to select naturally occurring aggregates for concrete, if these are available; they are usually specified in two batches, fine aggregate (sand) and coarse aggregate, and the grading of these must normally conform to the limits which are specified in the relevant standard (e.g. BS 882). Where a crushed rock type is used this too must be correctly graded. Crushed rock aggregate will normally require a higher water content to achieve a given workability and will therefore produce a weaker concrete than a naturally occurring aggregate. The workability of a concrete which is made with crushed rock can be improved by increasing the proportion of fine particles present. This allows the use of a lower water content, which improves the strength, but it results in a higher cement content being required, which increases the cost of the concrete.

The mix proportions of a concrete are specified in terms of the weights of materials which

125

are required to produce a unit volume of fully compacted fresh concrete and the principal factors which affect the mix proportions are the required strength and the required workability. The required strength is determined during the structural design calculations; the required workability depends on the nature of the structure, specifically on the sizes of the elements, the complexity of the pattern of reinforcement and the type of equipment which will be available to assist with the compaction. Mix design involves following a set of procedures in which the various properties which are required of the fresh and hardened concrete are used to derive systematically the specification for the concrete: a suitable water–cement ratio, cement–aggregate ratio and aggregate grading. The object is to obtain optimum mix proportions for the requirements of a particular structure. The process is a fairly complicated one and will not be described in detail here.

4.3.3.4 Reinforcement

The reinforcement which is used in concrete is normally in the form of steel bars, either of plain circular cross-section or with various surface treatments which increase the bond with the concrete (Fig. 4.30). The preferred diameters are 6, 8, 10, 12, 16, 20, 25, 32 and 40 mm, and the normal maximum length is 12 m. Reinforcement for slabs is produced in the form of square mesh, in individual pieces or in rolls, in which the bars are welded together at their crossing points. Meshes with most combinations of bar sizes and spacings can be readily obtained.

   Reinforcement is produced in both mild steel and high-yield steel. The latter allows much higher tensile stress to be specified for the reinforcement; its use is limited, however, because the critical factor which determines the amount of stress which can be permitted in reinforcement is frequently the need to control the amount of strain which occurs so as to prevent cracking of the concrete. The full tensile strength of high-yield steel cannot therefore be used in the design of reinforced concrete structures. The reinforcement which is used in pre-stressed concrete is hard-drawn

Square twisted bar

Ribbed and twisted bar

Stretched and twisted ribbed bar

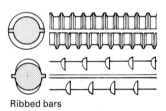
Ribbed bars

Fig. 4.30   Reinforcing bars are produced with a number of surface treatments to improve the bond with concrete.

steel wire; the preferred sizes are 4, 5 and 7 mm.

4.3.4 The behaviour of reinforced concrete

Reinforced concrete is a composite material whose structural action is both subtle and complex. The basic relationship between concrete and reinforcement is, however, fairly straightforward and is explained here in relation to elements which are subjected to bending-type loads.

   The effect of bending on any structural element is to place part of its cross-section in compression and part of it in tension. In the simplest types of reinforced concrete element, the compression on one side of the cross-section is resisted by the concrete and the tension on the other side by the reinforcement.[11] To appreciate the way in which the two materials interact it is necessary to visualise the distribution of stress which occurs within a beam when a bending-type load is applied.

The distribution of stress within a beam of any material depends on the configuration of the applied load. The action of a uniformly distributed load is to bend the beam downwards and cause tensile stress to be set up in the lower half of the cross-section and compressive stress in the upper half. The exact distribution of the stress is complex but a way of visualising it is shown in Fig. 4.31.

In Fig. 4.31a a series of circles has been drawn on the side of the beam. Following the application of the load these change their shape to become ellipses (Fig. 4.31b). The minor and major axes of the ellipses coincide with the directions of maximum compression and tension at each location. This allows the directions of the maximum tensile and compressive stresses (the principal stresses) to be plotted (Fig. 4.31c). Thus, at mid-span the direction of stress is parallel to that of the beam. Towards the ends of the beam the direction of the tensile and compressive stresses become progressively more inclined to the axis of the beam and the tensile and compressive stress lines cross each other. The material in these regions is stressed simultaneously in tension and compression in two orthogonal directions.

The diagram of principal stresses (Fig. 4.31c) shows only the directions of maximum tension and compression. The magnitudes of the stresses vary along each line from a maximum at the mid-span position, where they are parallel to the axis of the beam, to zero where the curved portions of the lines approach the top or bottom of the beam. The magnitudes of the stresses also vary between lines. They are greatest in the lines which run closest to the

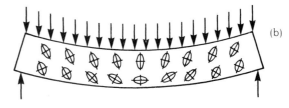

(a)

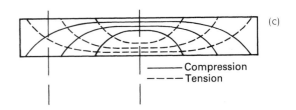

(b)

(c)

Compression
———— Tension

Fig. 4.31 The pattern of stresses in a beam. Circles drawn on the side of an unloaded beam (a) become ellipses as a result of the strain caused by the load (b). This gives an indication of the directions of maximum tension and compression at every location and allows principal stress lines to be drawn (c). The magnitude of the stress varies along each principal stress line from zero at the point at which it crosses the top or bottom surface of the beam to a maximum at the mid-span position. The magnitude of stress also varies between different lines with those closest to the top and bottom surfaces at the mid-span position carrying the greatest stress.

top and bottom of the beam at mid-span and decrease towards the interior of the beam.

The highest levels of stress occur on the cross-section at the mid-span point. The stress is tensile in the lower half of the cross-section and compressive in the upper half. Its magnitude varies from a maximum tensile stress at the lower extreme fibre to maximum compressive stress at the upper extreme fibre with zero stress half-way up at what is called the neutral axis.

---

11 In complex reinforced concrete elements compression reinforcement is also provided.

127

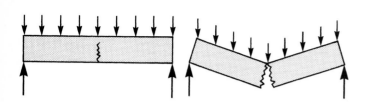

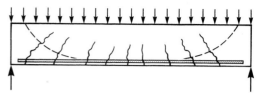

Fig. 4.32 Failure of an unreinforced concrete beam. The unreinforced beam fails as a result of the formation of a crack situated at the location of the highest tensile stress and at right angles to the direction of maximum tension.

Fig. 4.33 Crack pattern in a reinforced concrete beam. The reinforced beam carries significantly more load than the unreinforced beam. Tensile cracks form, always at right angles to the direction of maximum tension (given by the principal stress lines) in those locations at which the tensile strength of the concrete is exceeded. The beam continues to carry load, however, so long as each crack is effectively crossed by the reinforcement.

Figure 4.32 shows the behaviour of an unreinforced concrete beam which is subjected to increasing load. The symmetrical pattern of tensile and compressive stresses shown in Fig. 4.31c becomes established when the load is applied, with the maximum stresses occurring at the mid-span position. As the load increases, the magnitudes of the stresses increase and eventually the tensile strength of the concrete is exceeded at the lower extreme fibre (the location of the highest level of tensile stress). A crack forms at right angles to the direction of maximum tension and this rapidly propagates up through the beam which breaks into two halves. Because the compressive strength of the concrete is at least ten times greater than its tensile strength, the concrete in the top of the beam is relatively understressed when the tensile failure occurs.

The failure can be prevented if steel reinforcement is placed close to the lower surface of the beam (shown diagrammatically in Fig. 4.33). In the reinforced beam the tensile failure of the concrete at the mid-span position still occurs but the beam as a whole does not fail because the crack which forms is now crossed by the reinforcement. The reinforcing bar carries the tensile load and the beam continues to function so long as the cross-sectional area of the steel bar is sufficient to prevent it from becoming over-stressed in tension and it does not pull out of the concrete on either side of the crack. The frictional stress between the concrete and the surface of the bar (bond stress), is therefore an important consideration in determining the ability of the two materials to act compositely; this is the reason that special surface textures are often applied to reinforcement (Fig. 4.30).

If more load is applied to the beam than was required to cause the first crack to appear, the stress everywhere will increase and the point on the principal tensile stress line at which the tensile strength of the concrete is exceeded progresses along the line, in both directions, from the mid-span position. More cracks form (Fig. 4.33), always at right angles to the direction of maximum tension, but failure of the beam as a whole does not occur so long as the cracks are effectively crossed by the steel bar and the compressive strength of the concrete is not exceeded in the top half of the beam. As the tensile failure points move towards the ends of the beam the cracks, which always form at right angles to the direction of maximum tension, become more and more inclined to the beam axis and are less effectively crossed by the reinforcement.

Failure eventually occurs due to the formation of a diagonal crack which is not crossed by the reinforcement (Fig. 4.34). This type of failure is called a shear failure because the degree of inclination of the principal stress lines causes a shearing action on the cross-section rather than a simple bending action. Shear failure can be prevented by shaping the reinforcing bar so that it conforms to the profile of the line of the principal tensile

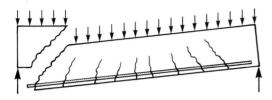

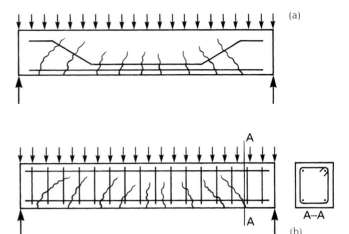

Fig. 4.34 Shear failure of a reinforced concrete beam. The beam depicted in Fig. 4.33 eventually fails due to the formation of an inclined crack at the end of the beam which is not effectively crossed by the reinforcement. This type of failure is called a shear failure.

Fig. 4.35 Beam with shear reinforcement. Bent up bars or links prevent shear failure by ensuring that all possible tensile cracks are effectively crossed by reinforcement.

stress. This is, however, difficult to carry out in practice and the solution which is normally adopted is either to bend up the main reinforcement bars in straight sections towards the ends of the beam (Fig. 4.35a) or to form a cage of reinforcement with links of steel (Fig. 4.35b). Both of these methods ensure that every potential tensile crack in the concrete is crossed by some reinforcing bars.

Figure 4.36 shows diagrammatically a typical pattern of reinforcement for a reinforced concrete beam. This consists of primary reinforcement in the tensile half of the beam to carry the tensile component of the bending stress, and shear reinforcement to carry the shear load caused by the inclination of the principal stresses. Secondary reinforcement is also normally provided (in this case longitudinal bars in the top half of the beam) to hold the primary reinforcement in position while the concrete is being placed around it. This secondary reinforcement also helps to control shrinkage and to carry secondary tensile stress which could occur, for example, due to thermal movement.

The behaviour described above, by which bending-type load is resisted by the composite action of concrete and steel, illustrates the way in which reinforced concrete functions. In large complex structures, such as multi-storey frameworks, the direction of bending within an individual beam or slab may vary due to the combined effects of structural continuity and load configuration. The distribution of the main reinforcement within the structure must

be such as to allow the effective resistance of all tension which occurs within the concrete. The pattern of reinforcement in real structures is therefore often considerably more complex than that shown in Fig. 4.36.

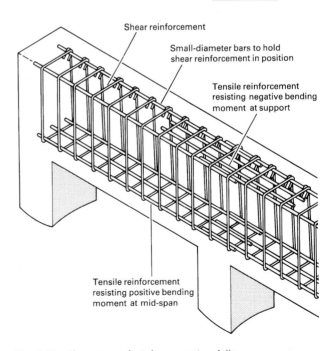

Shear reinforcement

Small-diameter bars to hold shear reinforcement in position

Tensile reinforcement resisting negative bending moment at support

Tensile reinforcement resisting positive bending moment at mid-span

Fig. 4.36 The pattern of reinforcement in a fully reinforced concrete beam.

4.4 Structural forms for reinforced concrete

4.4.1 Basic elements

Beams

Reinforced concrete beams are normally either rectangular in cross-section or combined with the slab which they support to form T- and L-shaped cross-sections (Fig. 4.37). The normal span range for reinforced concrete beams is 4.5 to 10 m. Spans of up to 20 m are occasionally used but depths of around 1.5 m and upwards are required for this, and a large volume of concrete is therefore involved. Spans greater than 20 m are possible but other types of structure will normally perform better in this span range. The depth which is required for beams depends on the span and the load which is carried. It is frequently determined from deflection rather than from strength requirements and in the normal span range a depth of around one twentieth of the span is required for a simply supported beam and one twenty-sixth of the span for a beam which is continuous across a number of supports. The breadth of a rectangular beam is usually around one third to one half of its depth.

Columns

Columns are constructed with a range of cross-sectional shapes, the most common being square, rectangular and circular. The primary reinforcement in columns is longitudinal reinforcement which contributes to the resistance of the compressive load and therefore reduces the required size of the cross-section. Transverse reinforcement, in the form of links, is also provided to give lateral support for the longitudinal reinforcement to prevent a bursting-type compression failure (Fig. 4.38). As in all compression elements, the principal factor which determines the dimensions which must

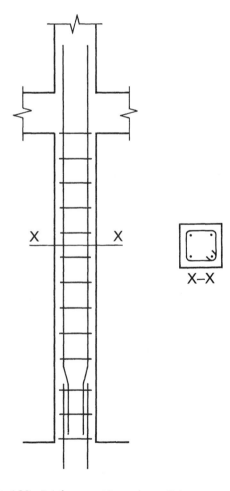

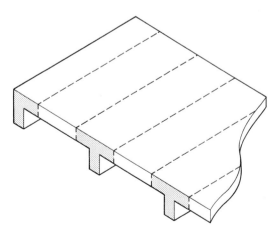

Fig. 4.37 T- and L-beams. Where reinforced concrete beams support a reinforced concrete slab the elements are normally cast together and act compositely to form T- and L-beams.

Fig. 4.38 Reinforcement in a column. Columns are provided with longitudinal reinforcement, to increase their compressive strength, and links to prevent buckling failure of the very slender reinforcing elements.

be adopted for columns is the need to avoid high slenderness, which would make the column susceptible to a buckling type of failure. The slenderness ratio of a concrete column is its effective length[12] divided by its least width and the British Standard (BS 8110, 'Structural Use of Concrete') recommends a maximum value of 60; 30 should be regarded as a practical maximum for most forms of construction, however, and to achieve reasonable load-carrying capacities, slenderness ratios in the range of 20–25 are normally adopted. Reinforced concrete walls perform in a similar way to columns and are made to conform to the same slenderness ratio requirements. In the case of walls the slenderness ratio is the effective height of the wall divided by its thickness.

Slabs

Reinforced concrete slabs can be either one-way- or two-way-spanning structures. The behaviour of a one-way-spanning slab is similar to that of a rectangular beam, and these elements can in fact be regarded simply as very wide beams. The primary reinforcement consists of longitudinal bars placed parallel to the direction of span on the tension side of the cross-section (that is the lower part in the mid-span position of slabs and the upper part near the supports where slabs are continuous over a number of supports).

Secondary reinforcement, in the form of straight bars which run at right angles to the direction of the span, is provided to control shrinkage and to distribute the effects of concentrated load (Fig. 4.39). The reinforcement for a solid slab normally takes the form of a mesh in which the bars of the primary and secondary reinforcement are welded together at the points where they cross. Shear stresses in slabs are very low, except close to columns, and primary reinforcement is not normally provided to resist shear.

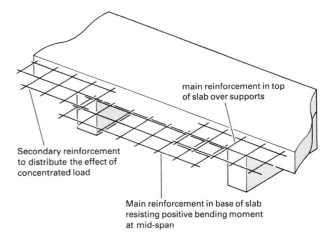

Fig. 4.39 Reinforcement in a slab. Slabs are provided with mesh reinforcement close to the top and bottom surfaces. If the bars are of equal thickness and spacing in both directions the slab will be capable of spanning simultaneously in both directions (two-way-spanning slab). If the degree of reinforcement in one of the directions is greater than in the other (larger bars at closer spacing) then the slab will span more effectively in that direction (one-way-spanning slab). In the latter case reinforcement is nevertheless provided in the non-span direction to distribute concentrated load and to control shrinkage of the concrete.

The depth of one-way-spanning slabs is normally approximately one thirtieth of the span for a simply supported slab and one thirty-fifth of the span for a slab which is continuous over a number of supports. The economic span for a solid slab is in the range 4 m to 8 m but this can be extended by using a ribbed form in which a proportion of the concrete in the lower half of the cross-section is removed. This type of slab is economic in the range 6 m to 12 m. If pre-stressing is applied the maximum spans are increased to 13 m for solid slabs and 18 m for ribbed slabs.

A two-way-spanning slab spans simultaneously in two directions and must be supported on beams, walls or rows of columns around its perimeter or at its corners. It should ideally have a square plan and the primary reinforcement is a square mesh of longitudinal bars; no secondary reinforcement is required. Two-way slabs are statically indeterminate structures and allow a more efficient use of material than one-way slabs. Similar span-to-depth ratios are

12 Effective length is based on the distance between points at which the column is supported laterally by other parts of the structure. Lateral support normally occurs at each storey level.

Fig. 4.40 Coffered slab. The coffered slab is a two-way-spanning flat-slab structure (i.e. a slab which is supported directly on columns). The coffers improve the efficiency by removing concrete from the tensile areas where it contributes little to the structural performance [Photo: Pat Hunt].

used but the maximum economic spans are higher; solid slabs are used for spans of up to 8 m and this can be extended to 16 m if the weight is reduced by removing some of the concrete on the tension side of the cross-section, and to 20 m if pre-stressing is used. A system of intersecting ribs, which form a square grid, is then created on the soffit of the slab; the resulting form is said to be a coffered or 'waffle' slab (Fig. 4.40).

Stairs

Stairs are designed as reinforced concrete slabs whose depths are equal to the waist thickness of the stair, and which are normally an integral part of the structure (Fig. 4.41). The requirements in respect of the ratio of span to depth which is specified are the same as for one-way-spanning slabs.

Precast components

The precasting technique, in which concrete components are manufactured in a factory and transported subsequently for erection on site, allows higher quality control to be achieved than is possible with *in situ* concrete. Among the advantages which this offers are higher concrete strength for given mix proportions, better quality of finish, greater dimensional accuracy and the economies of scale which are associated with the factory process. The factory method of manufacture also allows more complicated shapes of cross-section to be achieved than are possible with *in situ* concrete. This, in turn, makes possible higher levels of structural efficiency and can also facilitate the use of structural elements, such as beams and columns, as ducts for services. Precast concrete has therefore been widely

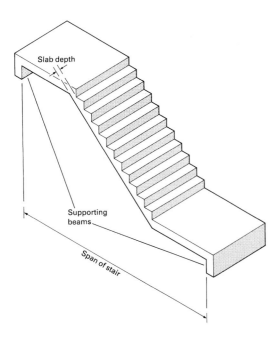

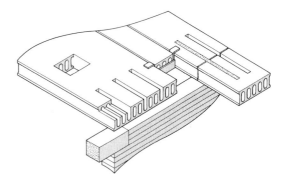

Fig. 4.42 Precast concrete flooring units. Precasting of concrete allows higher levels of quality control than are possible with *in situ* casting, which allows greater strength to be achieved from a particular set of mix proportions. This permits the use of reduced thicknesses for a given load-carrying capacity and the use of more complex cross-sections. The efficiency of the units depicted here is improved by the introduction of voids into the core.

Fig. 4.41 Reinforced concrete stairs are simply one-way-spanning slabs which span between the landings.

used in the context of heavily serviced buildings where the combination of structure and services is desirable.

A wide variety of precast concrete proprietary components is currently available ranging from simple rectangular-cross-section beams or slabs (Fig. 4.42) to complete framing systems for buildings (Fig. 4.43). In cases where no proprietary system is available, bespoke elements are used provided the scale of the project is such as to justify the setting up of the necessary production process.

Among the disadvantages of precasting are the difficulties associated with connecting elements together satisfactorily. Precasting also favours the adoption of building forms which are regular and repetitive because this simplifies the erection process and allows maximum advantage to be taken of economies associated with the mass production of identical units. The restrictions on form which are associated with steel frameworks are therefore also a feature of precast concrete structures.

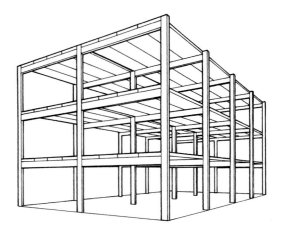

Fig. 4.43 Precast concrete framework. The precasting technique can be applied to all types of structural element and makes possible the construction of complete precast frameworks.

Precast components are frequently combined with *in situ* concrete and this allows the advantages of both forms of construction to be enjoyed (Fig. 4.44). The use of *in situ* concrete at the junctions of precast frameworks can solve the problems associated with jointing the units together, particularly the 'lack-of-

133

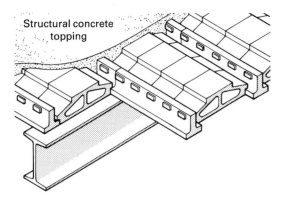

Fig. 4.44 Composite *in situ* and precast concrete. Precast units of complex shape are used here in conjunction with *in situ* concrete to form a composite floor deck with an efficient 'improved' cross-section.

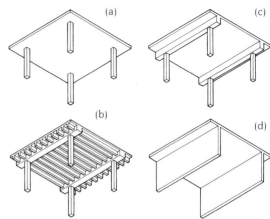

Fig. 4.45 Four basic types of reinforced concrete structure.
(a) Two-way-spanning ribless flat-slab.
(b) One-way-spanning ribbed flat-slab.
(c) One-way-spanning slab supported on rigid frame.
(d) One-way-spanning slab supported on loadbearing walls.

fit' problem and the provision of structural continuity. *In situ* concrete can also be used to allow more complex or irregular overall forms to be adopted.

4.4.2 Structural forms – cast-*in-situ* forms

4.4.2.1 Introduction
Reinforced concrete cast-*in-situ* forms are used principally for multi-storey buildings in which a frame-type structure is required. The form of the structure is determined by the same factors as influence the design of all structures, which are the need to provide effective resistance to

both gravitational and lateral load, and to achieve reasonable economy in the use of material. Although the mouldability of concrete theoretically gives the designer a wide choice of frame geometries, the need to minimise costs normally favours the use of rectilinear arrangements which require simple patterns of reinforcement and formwork. The variety of multi-storey forms which are used in

Table 4.1   Span ranges and size requirements for the four basic types of *in situ* reinforced concrete structure

Structure type	Span range (m)		Span/Depth		Column/wall
	Reinforced	Pre-stressed	Reinforced	Pre-stressed	slenderness
One-way-span slab	4×6 to 8×11	8×11 to 8×14	25	36	15 to 20
Two-way-span slab: solid	4×4 to 6×6	6×6 to 10×10	25 to 30	30 to 35	15 to 20
coffered	6×6 to 18×18	8×8 to 20×20	35	30	15 to 20
Beam/Column frame:					
one-way-span slab	3×6 to 6×12	6×12 to 8×15	slab 36	36	15 to 20
			beam 15 to 20	20 to 25	
two-way-span slab	4×4 to 8×8	8×8 to 15×15	slab 36	36	15 to 20
			beam 15 to 20	20 to 25	
Loadbearing wall:					
one-way-span slab	3 to 12		36		15 to 20
two-way-span slab	3×3 to 12×12		36		

practice is nevertheless considerable and they are categorised here into the four basic types of one-way-spanning flat-slab structures, two-way-spanning flat-slab structures, beam-and-slab structures and loadbearing-wall structures (Fig. 4.45).

All four basic types are beam-and-post arrangements and can be considered to consist of a floor deck system supported on a vertical structure of either columns or walls. Their properties, span capabilities and sizing requirements are summarised in Table 4.1.

In all cases, the design of the floor deck is determined principally by the requirements of gravitational load and this in turn dictates the pattern of the column or wall grid which is provided to support it. One-way-spanning floor systems are best carried on rectangular grids while two-way systems perform better on column or wall grids which are square. In either case the grid is normally kept as regular as possible for reasons of economy but it need not be perfectly regular.

The principal effect of lateral loads is upon the design of bracing systems. In many cases this will have no influence on the overall form of the structure because the beam–column arrangement will be self-bracing due to the high level of structural continuity which is possible with reinforced concrete. Some reinforced concrete structures require bracing walls for stability, however, and where these are necessary the internal planning of the building is affected.

Only the most basic, regular forms of each type of structure are described here to give an indication of the general arrangements and span ranges for which they are suitable. Often these basic forms are manipulated and distorted to produce more complex structural geometries (see Section 2.5).

4.4.2.2 One-way-spanning flat-slab structures
In this system the floor slab spans one way between rows of columns (Fig. 4.46). The arrangement is also known as ribbed-slab because the slab is frequently given a ribbed cross-section in order to improve its efficiency by removing concrete from the tensile side of

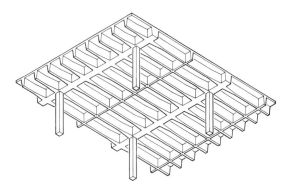

Fig. 4.46 One-way-spanning flat-slab system. The ribbed version depicted is used at the long span end of the span range for improved efficiency. For short spans, one-way-spanning flat-slabs have a simple rectangular cross-section.

the cross-section (Fig. 4.46). The ribs follow the span direction of the slab. The voids between the ribs are stopped short of the strip of slab between the columns and this solid area acts as a beam spanning between the columns. As this is a one-way-spanning system the column grid is rectangular with the slab spanning parallel to the long side of the rectangle. The normal span range is 4 m $\times$ 6 m to 8 m $\times$ 11 m but the maximum span can be increased to 8 m $\times$ 14 m if pre-stressing is used. The span/depth ratio is normally around 25 but can be as high as 36 in pre-stressed versions. Ribbed-slab structures are braced either by rigid frame action or by walls acting as diaphragm bracing in the vertical plane.

The particular advantages of the ribbed-slab system are the relative simplicity of the construction and high structural efficiency, which allows relatively long spans to be achieved with a small volume of concrete and a small depth of structure.

4.4.2.3 Two-way-spanning flat-slab structures
In two-way-spanning slab structures the floor deck system consists of a two-way-spanning flat plate of reinforced concrete which is supported directly on a grid of columns (Fig. 4.47). The column grid is normally square to maximise the two-way-spanning action. The

135

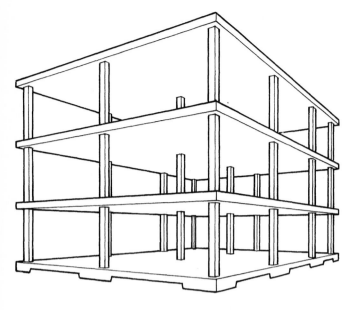

Fig. 4.47 Two-way-spanning flat-slab system.

Table 4.2 Overall slab thickness and column widths for two-way-spanning flat-slab structures

Column spacing (square column grid) (m)	Slab thickness (mm)		Column width (mm)
	Solid	Coffered	
4	150	–	200
5	175	–	200
6	200	300	250
7	250	300	250
8	275	300	250
9	300	400	300
10	–	400	300
12	–	500	400
14	–	500	500
16	–	600	600
18	–	700	700

structure is highly statically indeterminate and the resulting structural continuity allows the necessary strength and rigidity to be achieved with great efficiency. The economic span range for a solid slab is 4.5 m to 6 m which is increased to 10 m if pre-stressing is used. The span/depth ratio is typically in the range 25 to 30 for reinforced slabs and 30 to 35 for pre-stressed slabs. An indication of typical slab depths is given in Table 4.2.

The critical stresses in solid slabs are shear stresses which are very high in the vicinity of the supporting columns. The strength of the floor can be increased by thickening the slab locally at these points by a system of 'drop' panels or by using 'mushroom-type' column caps (Fig. 4.48); alternatively the slab can be strengthened in shear in the vicinity of the columns by the use of extra reinforcement in the form of steel shearheads (Fig. 4.49). These variations allow heavier loads to be carried or higher span-to-depth ratios to be achieved. The use of shear heads also allows voids for services ducts to be located close to the columns.

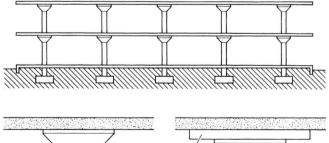

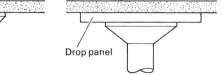

Drop panel

Fig. 4.48 Column caps and drop panels. The column cap and drop panel can be used independently or together to increase the shear strength of flat-slabs in the vicinity of columns where high shear forces require a greater thickness of slab.

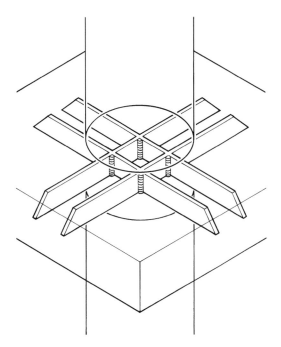

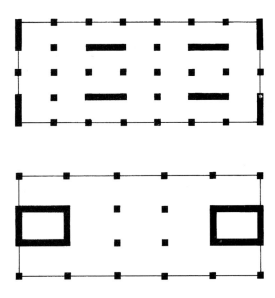

Fig. 4.50 Typical arrangements of vertical-plane bracing for flat-slab structures, which should be as symmetrical as possible, are shown here.

Fig. 4.49 Steel shearheads, which are cross-shaped arrangements of steel plates joined by welding, are an alternative method to column caps or drop panels for increasing the shear strength of flat-slabs in the vicinity of columns.

The maximum size of column grid of this type of structure can be increased to around 18 m for reinforced slabs and 20 m for pre-stressed slabs if a coffered system is used to reduce the weight of concrete present in the underside of the slab. The span/depth ratios for coffered slabs are around 25 for reinforced slabs and 30 for pre-stressed slabs.

The two-way-spanning slab system is dependent, to a large extent, on structural continuity for its strength. It performs best (i.e. allows the highest ratios of span-to-depth) with column grids which have at least three bays in each direction and in which the variation in the sizes of the bays is kept to a minimum. The efficiency of the system is such, however, that slab thicknesses are approximately the same as those in frame structures of equivalent span, i.e. a system in which the slab is supported on downstand beams.

Where a thin, solid slab is used its stiffness can be insufficient to allow rigid-frame action to develop with the columns in response to lateral load. Extra bracing must therefore be provided and *in situ* concrete walls are usually incorporated into the structure for this purpose. These must be arranged in two mutually perpendicular directions and can usually be accommodated around lift or stair towers (Fig. 4.50). In the long-span part of the range, that is for spans greater than around 10 m, the depth of the coffered slab which is required will be greater than 350 mm and the slab is usually sufficiently stiff to allow rigid-frame action to develop between the floor and columns, in response to lateral load; in these cases no additional bracing is required.

The high degree of statical indeterminacy which is associated with the two-way-spanning slab allows greater flexibility in the column grid than is possible with frame structures. The small construction depth of flat-slab structures, compared to frame structures, also facilitates the easy accommodation of a services zone and will usually result in a lower overall storey height than is possible with a beam–column frame.

137

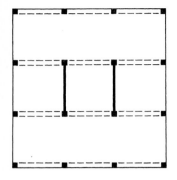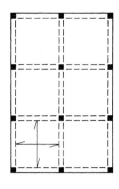

Fig. 4.51 Plan arrangements for reinforced concrete beam–column frames.
(a) One-way-span slab spanning between parallel beams.
(b) Two-way-span slab on a grid of beams.

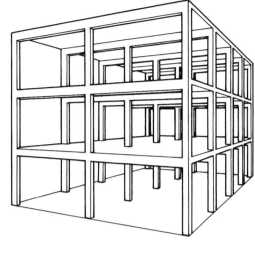

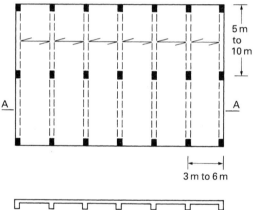

Fig. 4.52 Beam–column frame with one-way span slab. One-way span slabs are normally carried on parallel arrangements of beams supported by a rectangular column grid. No beams are provided in the direction of the slab span.

Due to the simplicity of the formwork and the high level of structural efficiency which is achieved as a result of the high degree of statical indeterminacy, two-way-spanning slab structures of both the plain and the coffered types provide very economical support systems for multi-storey buildings in which large wall-free areas are required. They are particularly suitable where the imposed load is high and uniformly distributed but are less suitable where concentrated loads are high, for example in buildings in which machinery has to be housed. They are also unsuitable for buildings in which the benefits of structural continuity are limited, either because large breaks in the continuity of floors occurs (Fig. 2.14) or because the building has a complicated geometry in plan and section (Fig. 4.15). In such cases a frame structure is normally required.

4.4.2.4 Frame structures

The distinctive characteristic of a frame is that it consists of an arrangement of beams and columns which supports slab floors. There are two basic types of reinforced concrete frame, the distinguishing feature being whether the floor is a one-way-spanning or a two-way-spanning system (Fig. 4.51).

In the frame with a one-way-spanning floor the floor-slab spans as a continuous system across a series of beams which are supported individually on columns. The beams act with the slab to form flanged T- or L-beams (Figs 4.37 and 4.52). Linking beams, running in the same direction as the slab, are not normally provided between the columns because the slabs themselves provide an adequate structural connection. The whole structure of beams, columns and slabs is cast in stages to form a continuous monolithic unit. Typical

Table 4.3 Span range and principal dimensions of reinforced concrete frame structures

Slab span	Beam span	Slab thickness	Beam depth (from top of slab)	Column width
(m)	(m)	(mm)	(mm)	(mm)
3	4.5	125	350	200
4	6.0	150	420	250
5	7.5	175	520	275
6	9.0	200	670	275
7	10.5	225	780	275
8	12.0	275	900	300
9	13.5	300	1060	300

dimensions for this type of arrangement are given in Table 4.3.

Frames of this kind can extend to a large number of bays in each direction and are best planned on a rectangular column grid with the slabs spanning parallel to the short side of the rectangle. The normal span range is 3.5 m × 6 m to 6 m × 12 m for reinforced concrete and this can be extended to 8 m x 15 m if pre-stressing is used. The most economic grid ratios vary between 1:1.5 at the short-span end of the range to 1:2 for longer spans. The span/depth ratio is around 36 for the slab and 15 to 20 for the beam.

So far as resistance to lateral load is concerned the one-way-spanning frame is rigid and self-bracing in the plane of the beam/column frames, due to the relatively high stiffness of the beams and columns and the rigid joints which exist between them, but is not stable in the direction of the span of the slab, because the stiffness of the slab is normally insufficient for effective rigid-frame action to be possible (Fig. 4.53). Additional bracing is therefore necessary and this is normally provided in the form of *in situ* concrete walls which act as vertical-plane diaphragm bracing (Fig. 4.51). These are constructed by simply extending pairs of columns to fill the space between them. As with other types of vertical-plane bracing it performs best if it is disposed around the building in a symmetrical arrangement and can usually be conveniently located at stairs and service ducts. The need to position the verti-cal-plane bracing correctly is a factor which affects the internal planning of buildings which have this type of structure.

Where a two-way-spanning slab is used in conjunction with a grid of beams it is necessary that the slab spans should be more-or-less the same in each direction. The column grid must therefore be more-or-less square (Fig. 4.51). The two-way-spanning floor structures have higher degrees of statical indeterminacy than one-way systems and this allows thinner slabs and shallower beams than the equivalent one-way system; it also increases the maximum economic span for a solid slab to about 8 m. Because frames with two-way-spanning floor systems have beams running in two mutually perpendicular directions they are completely self-bracing and do not require any

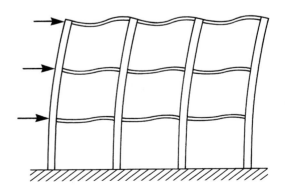

Fig. 4.53 Bracing of one-way-span frame systems by rigid joints is ineffective in the direction of the slab span due to the flexibility of the slabs.

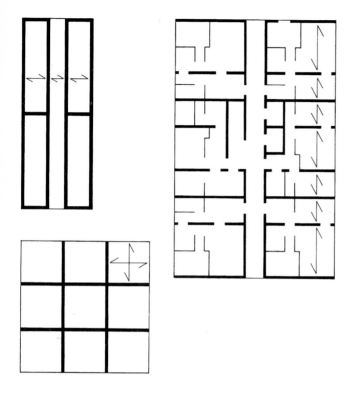

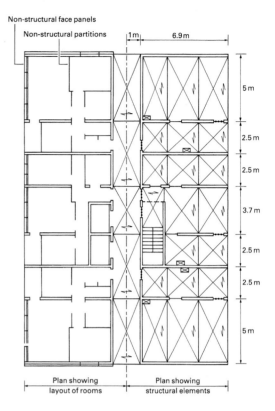

Fig. 4.54 Plan arrangements for loadbearing-wall structures of reinforced concrete.

Fig. 4.55 Plan arrangement for reinforced concrete
loadbearing-wall structure with one-way-spanning floor
slabs. Note that this conforms to the parallel-wall arrangement which is typical of loadbearing wall-structures in all
materials.

structural walls for stability. This is sometimes
the reason why a frame, as opposed to a flat-
slab structure is adopted.

The frame types which have been described
here are the most basic forms. Considerable
variation from these is possible although this
will normally increase the cost of the structure.
The simplest variation is the displacement of
individual columns from a strictly regular grid,
in order to accommodate some aspect of the
space-planning of the interior. If the displacement is kept within one quarter of the span it
can be accommodated easily by strengthening
the structure locally. Another common variation is a small change in the level of a floor
over a short area of the plan; this too can be
easily accommodated in *in situ* reinforced
concrete, as can the ramps and stairs which
are required for access. More significant variations from the standard forms are illustrated in
Figs 2.8, 2.10 and 2.14.

4.4.2.5 Loadbearing-wall structures

The planning principles for reinforced concrete
loadbearing-wall structures are similar to those
which are used for masonry structures
although the arrangement of walls is normally
simpler because the greater flexural strength of
concrete eliminates the need for local stiffening to combat buckling. Three plan-forms are
commonly used: cross-wall and spine-wall,
with one-way-spanning floors, and cellular,
with two-way-spanning floors (Fig. 4.54 and
4.55). The spacing between the walls is kept as
uniform as possible so that the slab spans are
equal and maximum benefit is obtained from
continuity. The most economic spans are
around 5 m to 6 m. In cross-wall and spine-
wall structures, in which the loadbearing walls

are parallel to one another, it is necessary to brace the structure with a limited number of walls running in the orthogonal direction; these can normally be located around stairs and service cores.

The advantages of reinforced concrete over masonry for this type of structure is that it allows the structural plan to be simpler and gives much more planning freedom generally: it is, for example, possible to omit some of the loadbearing walls at occasional locations and bridge the gap by reinforcing the wall above so that it acts as a deep beam.[13] The disadvantage of reinforced concrete is that it is almost invariably more expensive than the equivalent masonry structure.

4.4.2.6 Summary of cast-*in-situ* forms

In situ reinforced concrete structures perform best in circumstances in which imposed loads are high and they are therefore used principally to support multi-storey commercial and industrial buildings and only rarely as the structures for low-rise domestic-type buildings or single-storey buildings.

The main alternative to reinforced concrete for the multi-storey building is steel and the particular advantages of reinforced concrete are that it allows complex and irregular forms to be achieved more easily than with steel and that it is both durable and fire resistant, which eliminates the need for finishing materials.

4.4.3 Structural forms – precast forms

4.4.3.1 Introduction

In precast concrete construction the structural components are cast in a location which is different from that which they will finally occupy in the structure. They are normally manufactured in a precasting factory which is remote from the site but they may sometimes be cast on the site. The advantage of the latter is that it removes the constraints imposed by

the need to transport the components to the site. In either case the main advantage of precasting is the achievement of higher quality control, which results in higher strength, better durability and better surface quality than is possible with equivalent *in situ* concrete. It also reduces the time required to construct the building on site.

The particular advantages and disadvantages of precast concrete are summarised below.

Advantages
1 Almost all of the advantages of *in situ* reinforced concrete are obtained. The material has relatively high strength in compression, flexure and tension and is therefore suitable for all types of structural element and for skeleton-framework arrangements. It is also durable and fire resistant, which facilitates the exposure and expression of the structure.

2 The precasting of concrete allows better quality control to be achieved. The material is therefore stronger than equivalent *in situ* concrete and has a better standard of surface finish. Structural elements can therefore be more slender and are less likely to require the application of a finishing material to bring them to a satisfactory visual standard.

3 Precasting allows greater complexity of the form of individual components to be achieved. This characteristic can be exploited in several ways. In heavily serviced buildings, in which a large number of service ducts are required, it allows precast concrete columns and beams of complex cross-section to be used as frame elements and as service ducts. Precast concrete frames have therefore been fairly widely used for building types, such as hospitals and laboratories, in which the provision of services is a major factor in the design. The precasting technique has also been widely used for the manufacture of proprietary components with complex cross-sectional shapes.

---

13 The most famous building in which this technique was exploited was perhaps the Highpoint flats by Lubetkin and Arup.

4 Precasting of components leads to a much simpler site operation than is possible with *in situ* concrete. The building can therefore be erected more quickly with fewer temporary structures.

Disadvantages

1 One-off precast concrete systems tend to be more expensive than *in situ* equivalents. This favours their use for large-scale projects where economy of scale reduces the differential. It is claimed by the precast concrete industry, however, that the overall building cost is reduced if a precast structure is used instead of *in situ* concrete, due to the simpler and shorter site operation.

2 Inflexibility. Precast concrete units must normally be manufactured some time in advance of their being transported to the site and installed in a building to give time for the concrete to gain adequate strength. (Sometimes the highest load to which a component will be subjected will occur during transportation or installation.) Late changes to the design of the building cannot therefore be readily accommodated.

3 Standardisation. Another aspect of inflexibility is that, because precast structures are assembled on site from components which are fairly large, it is convenient to maintain the overall geometry of the structure in as simple a form as possible.

There is considerable incentive to standardise components so as to obtain the maximum re-use of moulds and to facilitate erection. Thus, while precasting offers the possibility that individual elements can have complex cross-sections and profiles, the overall form of the building must normally be relatively simple and repetitive. This is a disadvantage which precast concrete shares with steel.

4.4.3.2 Precast frame structures

Planning principles for precast concrete frames are the same as for *in situ* concrete frames: floors are normally of the one-way-spanning

type, in which case a rectangular beam–column grid is used but two-way-spanning floors, on a square column grid, are also possible. Normally the structure consists of individual beam, column and slab units which are erected on site in a similar way to a steel frame (Fig. 4.56); the joints between elements can be of the hinge or rigid type, depending on the detailing. If hinge-type joints are used (Fig. 4.57), additional bracing components are required and these can take the form of infill walls, either *in situ* or precast, or diagonal bracing. Where the beam–column joints are rigid the frames are self-bracing. Frequently in precast frame construction the joints between the individual units do not coincide with the beam–column junctions (Fig. 4.58). The units then have a fairly complex geometry which makes transport and stacking on site more difficult; the advantage is that it makes possible the achievement of a self-bracing frame without the need for site-made joints which are of the rigid type.

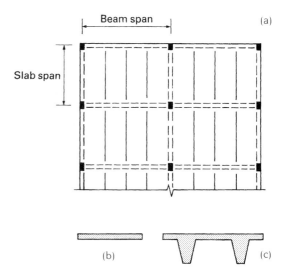

Fig. 4.56 Layout of a basic precast concrete framework.
(a) Typical plan arrangement in which a beam–column frame supports a one-way-spanning floor slab.
(b) and (c) Slab units can be of rectangular or ribbed cross-section depending on the span.

Table 4.4 Span range and principal dimensions of basic precast concrete frames

Slab span (m)	Beam span (m)	Slab depth (mm)	Beam depth (mm)
4	6.0	140	450
5	7.5	140	600
6	9.0	150	700
7	10.5	190	800
8	12.0	190	1000
9	13.5	190	1150
10	15.0	250	1300
11	16.5	250	1400
12	18.0	250	1500

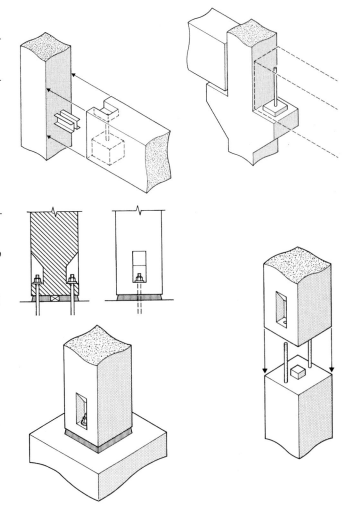

It is normal for precast concrete frameworks to have a regular, rectilinear form so as to maximise the standardisation of components but it is possible to adopt irregular grids. Typical element sizes for the normal span ranges of rectilinear frames are given in Table 4.4.

Columns are normally rectangular in cross-section but other shapes can be provided if this is necessary to accommodate irregular beam layouts or for architectural effect. The most favoured beam cross-section is the inverted T, as this facilitates the carrying of simple slab types. More complex shapes are used to accommodate changes in floor level.

Precast floor slabs are normally of the one-way-spanning type and are either solid, hollow-core or with a T-profile (Fig. 4.56c). All of these are ideally suited to a rectangular beam layout. Rhomboid plan-forms are fairly straightforward to produce but where the plan shape is highly irregular the section of floor involved is cast *in situ*.

Where vertical-plane bracing is provided by structural walls these can be precast units and can be used to provide support for floors as well as to resist lateral load. They are normally arranged as bracing cores around lift or stair wells. A reasonable number of bracing walls must be provided in two orthogonal directions and these should be arranged as symmetrically on plan as practicable.

4.4.3.3 Hybrid *in situ* and precast forms

Hybrid structures, in which both *in situ* and

Fig. 4.57 Joints in precast concrete frameworks. All of the joints shown are capable of transmitting shear and axial force only. They are therefore hinge-type joints.

precast concrete are used, allow the advantages of both forms of construction to be realised. The precast components will bring the benefits of factory production (high strength and efficiency, durability, good appearance, complex element cross-sections, dimensional accuracy, rapid erection) and the *in situ* parts allow complex or irregular overall forms and structural continuity between elements to be easily achieved.

The *in situ* and precast components in hybrid structures can be combined in two principal

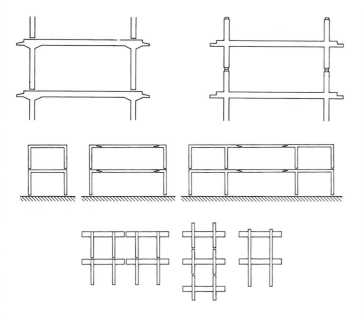

Fig. 4.58 Where rigid joints are required in precast arrangements, so that the structure is self-bracing, this is frequently accomplished by positioning the junctions between the elements at different locations from the beam-to-column connections.

Fig. 4.59 Hybrid *in situ*/precast structure. In this structure precast concrete perimeter units are used in conjunction with an *in situ* concrete framework. The superior standard of finish which precasting allows is exploited to allow the units to be exposed on the exterior of the building. [Photo: British Cement Association]

ways. In the first of these the structure consists of a mixture of element types each of which is either precast or *in situ* (Fig. 4.59). The ratio of precast to *in situ* components can vary widely. At one extreme an *in situ* framework of beams and columns, with rectangular shapes of cross-section, might be combined with precast stair and ribbed-slab elements whose more complicated profiles are more easily achieved under factory conditions. At the other extreme, the use of *in situ* concrete might be confined to the making of continuous joints in a structure in which all of the principal elements were precast.

A second type of hybrid structure is one in which the individual elements are formed by a combination of precast and *in situ* concrete acting compositely. With this type of arrangement the precast parts of the structure are invariably used as permanent formwork on which the *in situ* parts are cast (Fig. 4.60). Thus a ribbed slab may be formed from a slender precast soffit of complex shape on to which an *in situ* top is cast. Beams of complex shape can be formed in the same way. Composite hybrid structures are particularly well suited to the realisation of the advantages of both precast and *in situ* concrete.

The general arrangements of hybrid structures are similar to those for precast forms. They are either beam–column frameworks or flat-slab arrangements with column grids which are either square or rectangular depending, respectively, on whether the floor slabs are two-way or one-way-spanning systems.

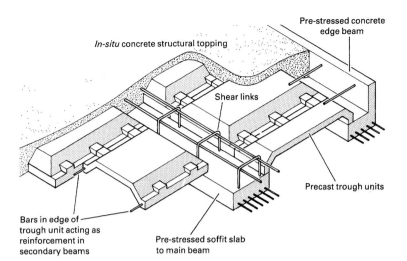

In-situ concrete structural topping

Pre-stressed concrete edge beam

Shear links

Precast trough units

Bars in edge of trough unit acting as reinforcement in secondary beams

Pre-stressed soffit slab to main beam

Fig. 4.60 In this hybrid structure precast beam and slab units act compositely with an *in situ* concrete topping slab.

4.4.4 Curved forms and structures of complex geometry

The mouldability and strength of concrete allow it to be cast into a wide variety of shapes and it has been extensively used to create envelopes which are based on curved forms (Fig. 4.61). The fact that these are more efficient structural forms than the post-and-beam structures which have already been described, either because they are close approximations to the form-active[14] shape for the loads which are applied to them or because they allow the proportion of under-stressed material which is present to be reduced in some other way, means that large spans can be achieved with very small volumes of structural material. Spans of up to 70 m are possible with shell thicknesses in the range 40 mm to 250 mm (Tables 4.5 and 4.6) and this represents a considerable saving compared with the volume of material which would be required to achieve the same span by using post-and-beam types of structure. A high degree of expertise is necessary, however, both in the design and in the construction of structures which have a complicated geometry and this tends to make them more expensive than forms with simpler geometries, despite the

large saving in material which is involved. Considerable reductions in the cost of the design and construction of shells is possible if some degree of regularity is introduced. For this reason most curvilinear or folded forms of structure are given a very simple basic geometry. The most favoured shell forms are those

Table 4.5 Typical thicknesses of hyperbolic paraboloid shells in reinforced concrete

Span (m)	Shell thickness (mm) At crown	At edges
10	40	50
20	40	75
30	40	100
40	75	130

Table 4.6 Typical thicknesses of elliptical paraboloid shells in reinforced concrete

Span (m)	Shell thickness (mm) At crown	At edges
10	40	50
20	50	80
30	60	120
40	70	170
50	80	200
60	100	220
70	130	250

14 See Appendix 1 for an explanation of the form-active concept.

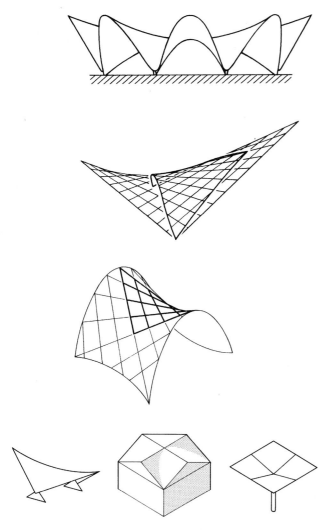

Fig. 4.61 Reinforced concrete lends itself to use in compressive form-active structural elements. The great efficiency of this type of structure allows the strength required for long spans to be achieved with very thin shells – typically between 50 mm and 150 mm. Various types of hyperbolic paraboloid shell are depicted here.

which can be described by an elementary mathematical relationship based on the Cartesian co-ordinate system. Examples are the cylinder, the sphere, the elliptical paraboloid, the hyperbolic paraboloid and the conoid (see Fig. 4.61 and also Joedicke, *Shell Architecture* Karl Kramer Verlag, Stuttgart, 1963).

Shell forms have other disadvantages besides the high cost of construction. They do not perform well when subjected to concentrated loads and their use must therefore normally be restricted to that of providing a free-standing envelope to which no other substantial components, such as services ducts or services machinery, are attached. The fact that the structural envelope of a shell-type building is very thin can produce a number of difficulties for the designer, which are not present when more conventional types of structure are used. If, as is usually the case, the shell is used as the sole external envelope an obvious deficiency is its poor thermal insulation and low thermal mass. Another disadvantage arises from the fact that it is not possible to accommodate the large number of systems which occur within a building, such as wires, pipes, etc., within the very small structural zone of a shell. This problem does not arise in buildings which have more inefficient post-and-beam type structures because the fairly large structural volume involved provides a zone within which many required components (ducts, cable runs) can be accommodated; where a very efficient structural form is used these components must be located elsewhere. The high complexity of the shell form, combined with its other disadvantages therefore make it suitable for a rather limited range of building types.

4.4.5 Conclusion

The advantages of reinforced concrete as a structural material stem from its physical properties and its mouldability. Its properties allow it to be used in both compressive and bending-type structural components and its strength is such that it can be used to make frameworks in which fairly large spans are involved. Of the non-structural physical properties, the good durability and fire resistance of concrete are probably the most important; these make it fit for use as a building support system without the need for extra finishing materials or components. The mouldability of concrete allows good structural continuity to be achieved, through the use of cast-*in-situ* rigid joints, and makes possible a considerable variety of geometries.

Chapter 5

Masonry structures

5.1 Introduction

Masonry has been used for the construction of buildings from earliest times and the history of its use comprises buildings in virtually every architectural style. It is a composite material in which shaped natural stone or manufactured bricks or blocks are bedded in mortar. It is brittle, with moderate compressive strength but little strength in tension and it therefore performs well in structural elements in which axial compression predominates, such as walls, piers, arches, domes and vaults, but will crack if subjected to tensile stress resulting from the application of bending. Where masonry elements carry bending-type loads they must be designed such that tensile stress is maintained at a very low level. If the element is subjected to pure bending, as in the case of a lintel or archi-trave, the tensile bending stress has to be minimised by the adoption of generous dimensions for the cross-section and an acceptance of very short spans. If the element is subjected to a combination of bending and axial compression the arrangement must be such that the tensile bending stress never exceeds the level of axial compressive stress. This is done by ensuring that the elements are sufficiently thick.[1]

Where the intensity of bending is high, as can occur in long or high stretches of wall subjected to wind loading or where high walls support vaults or domes, the overall thickness of element required to contain the bending

stress within reasonable limits can be large. In such cases various measures can be adopted to achieve high overall thickness without the need for a large volume of masonry. Often, the structural devices which make this possible provide the stimulus for architectural innov-ation (see Section 5.2).

The forms which are appropriate for masonry structures are suitable for any material with similar properties (i.e. materials with moderate compressive strength but little strength in tension or bending). Baked earth, mud and unreinforced concrete are examples of such materials.

5.2 The architecture of masonry – factors which affect the decision to use masonry as a structural material

5.2.1 The aesthetics of masonry

The architectural vocabulary of masonry is that of compression: the structural forms are those of the loadbearing wall, the buttress, the arch, the vault and the dome. The two great trad-itions of Western architecture, the Gothic and the Classical, were each developed through the medium of masonry structures and each found its greatest architectural expression in masonry. Masonry accounts for some of the most spectacular of the world's buildings, both in terms of physical scale and visual qualities.

The architectural use of masonry is here reviewed under the three headings of vaulted halls, domes and post-and-beam structures. Much of the architecture described is pre-twentieth century but this is necessary so that the full range of the architectural potential offered by masonry can be illustrated.

---

1 The factors on which the relationship between bending stress and the dimensions of structural elements depends are explained in Macdonald, *Structure and Architecture*, Chapter 2.

Fig. 5.1 Cathédrale Ste-Cécile, Albi, France, 13th century. This building, which has the widest vaulted nave of all the French Gothic cathedrals (18 m span), is constructed of brick-work masonry and demonstrates the scale and complexity of architectural form which is possible in this material. [Photo: A. Macdonald]

Vaulted Halls

Vaulted hall is the generic term for a type of building which consists of a large interior space constructed entirely of masonry; a space, in other words, which is roofed by a masonry vault supported on masonry walls. This group of structures encompasses some of the largest and most spectacular of masonry buildings, including the large basilicas and bath houses of Imperial Rome and the Gothic cathedrals of medieval Europe (Fig. 5.1).

Where a large enclosure is made entirely of masonry the use of a compressive form-active

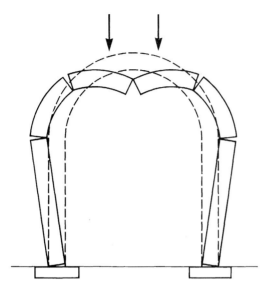

Fig. 5.2 Failure of a vault. Vaults exert horizontal force on their supports. If the supports are high walls, these must be sufficiently thick not to overturn and have sufficient bending strength not to be overstressed by the bending moment caused by the horizontal force at their top. If significant horizontal movement occurs at the tops of the walls the vault is likely to fail in the manner shown here.

structure,[2] such as a vault or a dome, is necessary to achieve the horizontal span without incurring high bending and therefore tensile stress. The action of a vault produces horizontal thrusts at the points of support, however, which, in the case of a vaulted hall, are the tops of the supporting walls (Fig. 5.2). The walls are therefore subjected to a combination of axial compression, due to the weight of the vault, and bending, due both to the lateral thrust of the vault and to any eccentricity which is present in the transfer of its weight to the wall. The configuration of the structure must be such that the axial compressive stress in the walls is always greater than the tensile component of the bending stress. This can only be achieved if the wall is of adequate thickness. The very thick walls required to

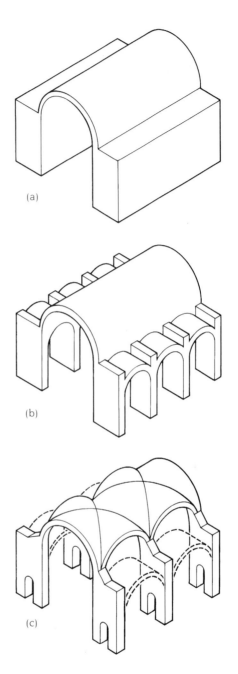

Fig. 5.3 Evolution of the form of the vaulted hall.
(a) The earliest vaulted halls relied on very thick walls for stability. The volume of masonry involved was very large.
(b) The use of the concept of the buttress allowed a degree of stability to be achieved which was equivalent to that provided by the solid wall but with a greatly reduced volume of masonry.
(c) The cross-vault concentrates the load at the buttresses and creates flat areas of wall above the springing point of the vault which may be used as fenestration.

2 See Appendix 1.

Fig. 5.4 Basilica Nova, Rome, 4th century CE. The architects and engineers of Imperial Rome exploited the architectural opportunities offered by the buttressed, cross-vaulted hall to create interiors of high architectural quality.

support high-level vaults do not have to be solid, however, and much of the architectural expression of large vaulted hall structures has been derived from the manner in which the voiding of the walls was carried out.

The earliest of the large vaulted halls were the basilicas and bath houses of Imperial Rome. The evolution of the form is shown diagrammatically in Fig. 5.3. The simplest form of the building consisted of a barrel vault supported on very thick, solid walls (Fig. 5.3a). The efficiency of this structural arrangement is low and is improved if voids are incorporated into the walls (Fig. 5.3b). The solid areas between the voids are, in effect, buttresses; the voids are also vaulted. A logical progression from this configuration is the adoption of a cross-vault arrangement (Fig. 5.3c) as this concentrates the load from the vault on to the solid areas of the walls. It also creates flat areas of wall above the original wallhead level (i.e. in the zone of the vault) in which clerestory lighting can be inserted.

The Basilica Nova (4th century CE) (Fig. 5.4) and the main building of the Baths of Caracalla (3rd century CE) are examples of this type of structure. In both cases the structural armature consisted of the vaulting system described above, constructed in mass concrete, encased in a thin skin of brickwork which served as permanent formwork (Fig. 5.5). In both of these examples this structural armature was clad in marble to create a sumptuous interior.

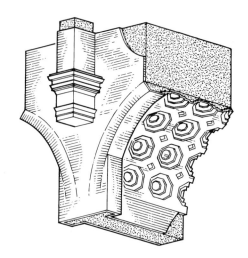

Fig. 5.5 Detail of Basilica Nova, Rome, 4th century CE. The largest interiors in Rome were constructed in unreinforced concrete, which was placed in a thin skin of brickwork which acted as permanent formwork. The structural carcass was then faced in marble to create a sumptuous interior. The system of construction allowed very large-scale buildings to be constructed economically.

The architectural opportunities offered by the structural vocabulary of masonry were therefore fully exploited by the Roman builders who used them to create sequences of interior spaces which varied in height, volume and lighting intensity. Buildings of this type played a significant role in the development of the classical language of architecture. The stylistic devices which evolved in response to the forms which were adopted out of structural necessity were to be much imitated at the time of the Italian Renaissance, and subsequently.

The Gothic cathedrals of the medieval period, in which vaulted roofs were balanced on highly complex buttressed wall systems, belong to the same generic type of building as the vaulted basilicas and bath houses of Imperial Rome. It is interesting to note, however, that, although the architects of the Gothic period were faced with very similar structural problems to those of their Roman predecessors and that their structural solution to it was similar, they produced an architecture which was quite different in style and character from that of the Romans.

The structural solution adopted by the Gothic builders was, therefore, similar structurally to that of the largest Roman buildings. The vaulted roof of a large hall (the nave of the church) was supported on vertical structural systems of great thickness but from which much mass was extracted to improve the efficiency with which material was used (Fig. 5.6). In the case of the Gothic buildings the sculptural effect occurred on the exterior of the building and took the form of buttresses, flying-buttresses, finials and all of the other elements of the Gothic architectural vocabulary. The basic structural arrangement was the same as in the Roman buildings but the architectural treatment of it was entirely different.

In some of the later buildings of the Gothic period the degree of refinement was extreme (Fig. 5.7). In these buildings the loads from the vaulting were concentrated into slender columns braced by stabilising buttresses. The walls linking these were extensively pierced with traceried windows. The walls, in fact, had a minimal structural function and had become virtually non-loadbearing curtain walls. The structural system of these late Gothic cathedrals was really that of the rigid frame rather than of the loadbearing wall – a remarkable achievement in a material with minimal tensile strength. The relationship between architectural aspiration and structural technology was

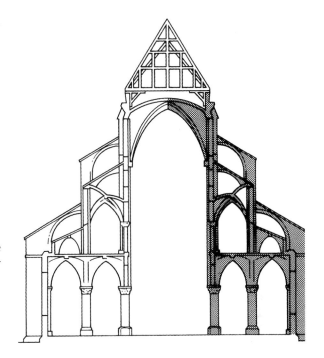

Fig. 5.6 Notre-Dame Cathedral, Paris, France, 12th century. As in the case of the Roman vaulted halls the vault here is supported on a wall system of great thickness. The architectural treatment of the buttressing system is quite different however.

perhaps closer in these buildings than has ever been achieved subsequently.

Present-day tall single-storey buildings, which are used for single-space large enclosures such as sports halls, normally have roofing systems which do not exert significant lateral thrust at the tops of the walls. They are not therefore true vaulted halls. The walls are nevertheless subjected to significant out-of-plane forces due to the action of wind pressure and must have adequate strength in bending to resist this. The strategies which are adopted in modern practice to achieve this are similar to those which were used in historic times – that of providing high overall thickness without the use of an excessive volume of masonry. The fin and diaphragm wall systems are examples of this (Fig. 5.41). A notable alternative method is the use of corrugations as is seen in the distinctive work of the Uruguayan architect Eladio Dieste (Fig. 5.8).

151

Fig. 5.7 Notre-Dame Cathedral, Chartres, France, 12th and 13th centuries. The main structural elements here are the quadripartite vaulting of the roof and the supporting piers which act in conjunction with external flying buttresses. The sections of wall between the piers can be regarded as non-structural cladding. The structural arrangement is therefore that of a skeleton framework – a remarkable achievement in a material such as masonry. [Photo: Courtauld Institute]

Fig. 5.8 Church at Atlantida, Uruguay, 1958. Eladio Dieste, architect. The plan-form of the walls of this building is that of a sinusoidal curve whose amplitude is greatest at the level where the bending moment is highest. The arrangement allows a large overall thickness of wall to be achieved with a minimal volume of masonry. This is a modern equivalent of the flying buttress. [Photo: Vicente del Amo]

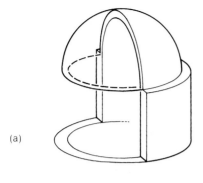

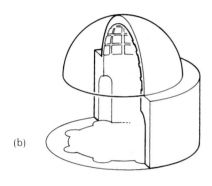

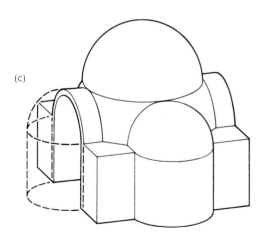

(a)

(b)

(c)

Domes

The dome is a similar structural form to the vault. In the history of the architectural treatment of the dome, a similar gradual progression to that which occurred with the vault may be observed, that is a progression from fairly simple, solid arrangements to gradually increased levels of refinement (Fig. 5.9). The structural action of the dome is slightly different from that of the vault, however, because although, in theory a perfectly intact dome exerts no horizontal thrust on the supporting structure, which can be a cylinder of masonry or even a ring of columns, in practice, horizontal thrusts do frequently occur at the supports. The precise behaviour depends on the profile of the dome.

The easiest type of dome to construct is the hemisphere because the constant radius of curvature allows simple formwork to be used. The hemisphere is not an ideal structural form, however, because the lower parts of hemispherical domes are subjected to horizontal tensile stresses in the circumferential direction (hoop stresses), which can give rise to meridional cracking (Fig. 5.10). The dome then

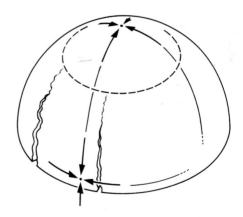

Fig. 5.9 Evolution of the form of the domed enclosure.
(a) The earliest domes were carried on very thick solid walls.
(b) At the Pantheon in Rome (2nd century CE) niches were provided in the thick walls and the underside of the dome was coffered to reduce the volume of material required. Both of these devices were exploited in the architectural treatment of the interior.
(c) The dome of the Hagia Sophia, in Istanbul (6th century CE), and those of several other large Byzantine buildings were carried on an arrangement of arches and pendentives which transmitted the weight to four massive piers.

Fig. 5.10 Internal forces in a hemispherical dome. A hemispherical dome which is intact exerts no lateral force on the supporting structure. The lower parts of hemispherical domes are subjected to horizontal, tensile 'hoop' stresses, however, which can cause cracks to develop as shown. If this occurs the individual segments of the dome behave as arches and do exert horizontal force on the supports. Most domes of masonry or unreinforced concrete exhibit cracking caused by hoop stress and do therefore require support structures which are capable of resisting horizontal load.

Fig. 5.11 The Pantheon, Rome, 2nd century CE. This was the largest domed structure of Roman antiquity. The coffering of the inner face of the dome and voiding of the walls suggest that the 'architects'/'engineers' responsible for this building had a good intuitive or empirical understanding of the structural behaviour involved. They used these features to excellent architectural effect. [Copyright: RIBA Photo Library]

behaves as a series of arches, arranged around a central axis, which exert a significant horizontal thrust on the supporting structure. The effect is lessened if the profile of the dome is pointed rather than hemispherical. Cracking, and the resultant horizontal thrusts, can be minimised if material capable of resisting the tensile hoop stresses is incorporated into the dome structure. Iron chains have been used for this purpose.

The earliest of the great domes of the Western architectural tradition was that of the Pantheon in Rome, constructed in the second century CE (Fig. 5.11). This is 43 m in diameter and was made of Roman concrete. The internal profile was hemispherical and the dome was supported on a cylindrical drum of concrete faced internally in marble and externally in brickwork. Meridional cracks developed in this dome, in the manner described above, but the resultant outward thrusts were absorbed by the very thick walls of the supporting drum. The supporting system of voided walls was of a

similar configuration to that of the systems which were used to support vaulted halls, which suggests that the Roman architects were well aware that the drum would be subjected to bending as a result of outward thrusts from the dome.

The dome of the Hagia Sophia in Istanbul (6th century CE) (Fig. 5.12) has a slightly smaller span than that of the Pantheon (31 m) but the design is considerably more audacious. This dome is much thinner than that of the Pantheon and it is supported on four piers, to which the weight is transferred by four arches and pendentives (in a square arrangement on plan) rather than a cylindrical drum of masonry. The piers are heavily buttressed by the surrounding parts of the building but nevertheless show signs that outward movement occurred due to the lateral thrusts produced by the dome and its supporting arches. The configuration of the structure suggests that the architects were aware of the crack pattern which had developed in the

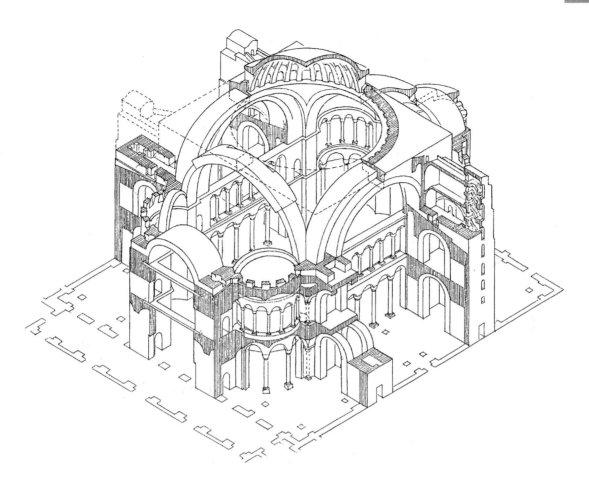

Fig. 5.12 Hagia Sophia, Istanbul, Turkey, from the sixth century CE. The dome which spans the central space is carried on four arches which transmit the weight to four massive piers. These are buttressed by the surrounding parts of the building. The combination of dome and arches allowed a complex interior of very large volume to be created. The architectural effect was one of a feeling of immense space. [Copyright: Rowland Mainstone]

Pantheon dome[3] and that they had a sound qualitative awareness of the behaviour of form-active masonry structures.

A significant development in the history of the masonry dome was the use of two structural skins separated by a voided core. The advantages of this were that it allowed great overall thickness to be achieved without the occurrence of a severe weight penalty, that the weathertightness of the structure was improved and that it allowed greater freedom in the choice of the shapes for the inner and outer visible surfaces of the dome.

The first very large structure to which this idea was applied was the fifteenth-century dome of Florence Cathedral (S. Maria del Fiore) by Brunelleschi (Fig. 5.13).[4] The dome was much higher in relation to its span than those of the Pantheon or the Hagia Sophia and this would have reduced the hoop stresses.

3 See Mark (ed.), *Architectural Technology up to the Scientific Revolution*, Cambridge, MA, 1993, Chapter 4.

4 An account of the construction of this very significant structure has been given elsewhere – see Mark (*op. cit.*) and Mainstone, *Developments in Structural Form*, London, 1975, Chapter 7.

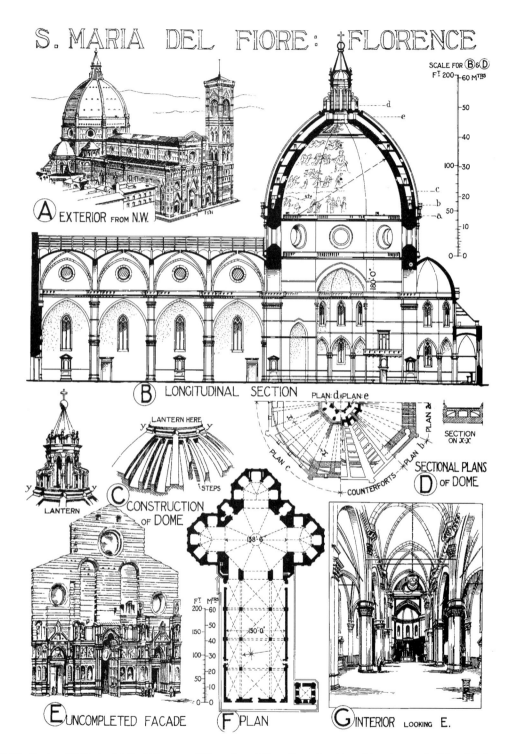

Fig. 5.13 Dome of S. Maria del Fiori, Florence, 15th century. Filippo Brunelleschi, architect/master builder. This brickwork dome has two skins separated by structural diaphragms. This significant innovation – one of several in this very important structure – allowed a measure of independence to be achieved between the internal and external profiles of the structure. [Copyright: RIBA Photo Library]

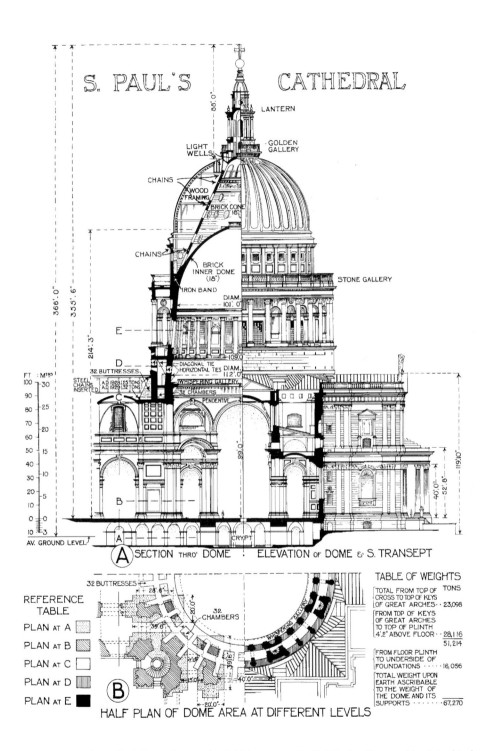

Fig. 5.14 Dome of St Paul's Cathedral, London, England, 17th century. Sir Christopher Wren, architect. In this dome the principle of twin structural skins, which was first used in Florence by Brunelleschi, was extended to allow complete independence to be achieved between the interior and exterior profiles of the dome. Here, the outer structural skin – a cone of brickwork – is not actually seen but is used to support a timber outer dome of non-structural cladding.
[Copyright: RIBA Photo Library]

157

It nevertheless suffered slight meridional cracking but this could have been due to thermal movement rather than primary load-carrying stress. The walls of the hexagonal drum which supports the dome are not especially thick and the lack of outward movement of these suggests that no significant outward thrust occurred at the base of the dome.

The dome of St Peter's in Rome (early seventeenth century) was designed by Michelangelo Buonarotti and modified by Giacomo della Porta. This too was a double-skin masonry dome but it was much flatter than the Florentine dome. Two iron chains were built into it to prevent meridional cracking but cracking which was extensive enough to threaten the stability of the structure nevertheless occurred. The dome was stabilised in the early eighteenth century, under the direction of Giovanni Poleni, by the insertion of five additional chains, at which time it was discovered that one of the original chains had broken. Poleni's measures were successful and no further intervention has been required.

The architectural possibilities offered by the double-skin dome were extended in the seventeenth century by the configuration adopted by Christopher Wren in his design for the dome of St Paul's Cathedral in London (Fig. 5.14). Wren was aware of the problems which had occurred at St Peter's and adopted a very lightweight double-skin structure of brickwork masonry. The inner skin formed the inner visible profile of the dome and was more-or-less hemispherical. The outer structural skin was a cone of brickwork which, with straight sides, rose from the same springing as the inner dome and provided direct support for the masonry lantern which formed the climax of the building's exterior. This outer structural skin was bound with chains but the profile of the cone was such that hoop stresses were minimised. A lightweight secondary structure of timber was built on top of the outer skin and supported the visible external profile of the dome. Thus did Wren create one of the most spectacular examples of the architectural scenic effect – one of the several examples of

ingenious architectural artifice which he employed at St Paul's.[5]

The brilliance of Wren's configuration was that it allowed a dome to be constructed from components of relatively light weight, which could be supported on an underlying system of slender columns and arches, but which nevertheless provided internal and external profiles which were compatible with their respective architectural schemes. A similar system was adopted by Jules Hardouin-Mansart for the Dome Church of Les Invalides in Paris. This type of arrangement was to be adopted for virtually all subsequent large masonry domes in the architecture of the West.

No significant developments in masonry dome construction have in fact occurred since the time of Wren and Mansart. The great domes of the eighteenth century had similar structural schemes and, in the nineteenth century, masonry was displaced by metal and reinforced concrete for long-span structures.

Post-and-beam structures
The most common architectural use of masonry has always been as the vertical elements in post-and-beam structures in which timber or reinforced concrete was used for the horizontally spanning elements. Buildings of this type have been constructed from the very beginning of the Western architectural tradition. The majority are rectangular in plan with flat or pitched roofs. The horizontal structural elements are carried on sets of loadbearing walls which are parallel to each other on plan but the plan, for structural stability, must also include some walls running at right angles to these. The resulting buildings are therefore multi-cellular. This generic plan and constructional arrangement for loadbearing-wall buildings has remained unchanged for centuries. The variety of architectural treatments which have been possible has nevertheless been very wide and the building type has been used in

---

5 A similar system of timber exterior supported on a masonry core – in this case a single skin – had been used at St Mark's in Venice in the eleventh century but Wren may not have known of this.

the context of most of the stylistic types and variations of the Western architectural tradition.

The classical language of architecture originated in a system of ornamentation for post-and-beam structures. Although the earliest versions may have been constructed in timber the primary sources of the classical orders were the masonry temples of Greek antiquity (Fig. 5.15). These were post-and-beam structures with vertical elements of masonry and horizontal elements of timber or masonry. The smallest of these buildings were single-cell structures but, due to the need to limit spans, larger versions were subdivided internally by parallel sets of walls or columns which conformed to the generic plan for the loadbearing-wall building described above.

Versions of the generic plan occurred in every period of Western architecture. A few notable examples serve to illustrate this. The villas of Andrea Palladio (Fig. 5.16), dating from the sixteenth century, show the normal pattern of parallel loadbearing walls as do the many subsequent large houses of Europe and North America whose forms were inspired by those of Palladio and other architects of the Italian Renaissance (Fig. 5.17). The much

Fig. 5.15 The Parthenon, Athens, 5th Century BCE. This is the most refined example of post-and-beam construction in the Western architectural tradition. The short spans involved are a consequence of the inherent weakness of masonry in tension and therefore in bending. [Copyright: RIBA Photo Library]

humbler tenement housing which was built in the nineteenth century to accommodate the growing urban populations of the cities of industrial Europe shows a different plan-form but one which nevertheless conforms to the basic arrangement of parallel loadbearing walls (Fig. 5.18). The variety of architectural treatments of masonry structures in terms of massing, ornamentation and fenestration is therefore very great despite the similarities exhibited by the plan-forms of loadbearing-wall buildings through the centuries.

In the twentieth century, the vast majority of masonry loadbearing-wall buildings have conformed to the traditional plan-form and the versatility of the medium is such that the vocabulary of Modernism, with, for example, large areas of wall glazing in the external walls, has been accommodated without difficulty.

The development of new constructional methods, especially the introduction of steel

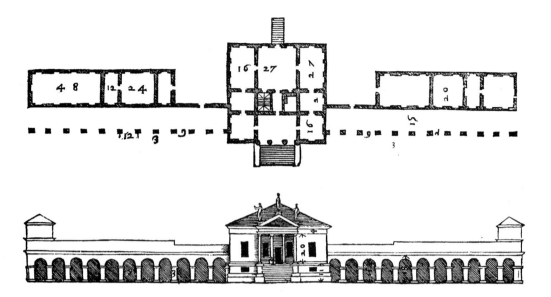

Fig. 5.16 Plan and elevation of the Villa Emo, Fanzolo, Italy, 1564. Andrea Palladio, architect. The vaulted basement and timber upper floor of this building are carried on the parallel cross-walls of a typical loadbearing masonry plan. [Illustration from Palladio's *Quattro Libri dell' Architettura* of 1570].

Fig. 5.17 Blenheim Palace, Oxfordshire, 18th century. John Vanbrugh and Nicholas Hawksmoor, architects. The plan of this very large house shows an arrangement of parallel walls which is typical of multi-storey loadbearing-wall construction.

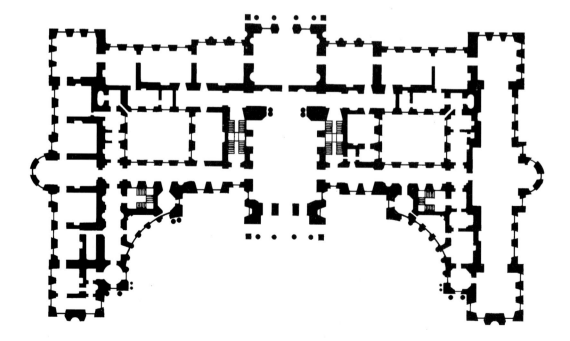

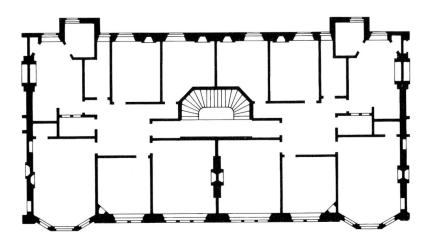

Fig. 5.18 Plan of a tenement building, Glasgow, 19th century. This much humbler building also shows the typical parallel-wall arrangement which is required to ensure that adequate support is provided for all areas of wall and roof.

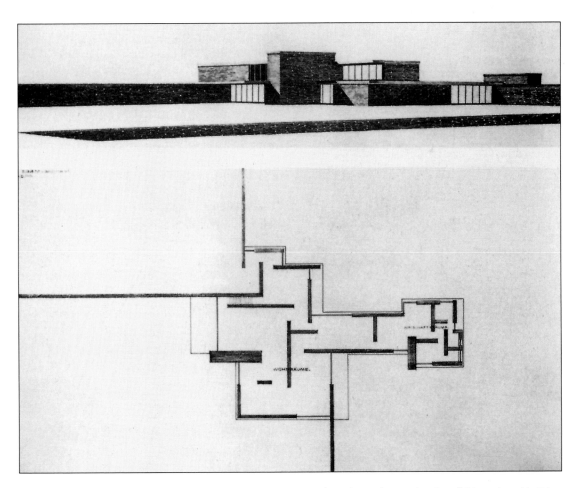

Fig. 5.19 Brick Country House Project, 1923–24. Ludwig Mies van der Rohe, architect. The plan of this projected building does not conform to the traditional parallel-wall arrangement seen above. Plans like this became possible only after systems of horizontal structure which were capable of spanning in more than one direction (reinforced concrete slab, steel space framework) had been developed.

161

and reinforced concrete, has made possible the creation of quite different plan-forms in masonry, however. The 'brick country house project' of 1923–24 by Mies van der Rohe may serve as an example (Fig. 5.19). Here an irregular pattern of loadbearing walls was adopted to accommodate the Modernist preoccupation with free-flowing space. This building was never actually constructed but if it had been, the organisation of the horizontal structure of the roof would have been necessarily more complex than was necessary with the generic parallel-wall type of plan. The choices for the horizontal structure system would have been either a two-way-spanning system (reinforced concrete-slab or steel space framework) or one based on a complex system of primary and secondary beams.

This project demonstrates that with modern construction techniques virtually any plan-form is possible in loadbearing masonry. Non-regular arrangements of walls produce a requirement for more complex systems of horizontal structure than those which are based on the traditional, and more sensible, parallel-wall type of plan.

5.2.2 The technical performance of masonry as a structural material

5.2.2.1 Introduction

The decision by an architect to use masonry as the structural material for a particular building is based on a knowledge of its properties and capabilities in relation to those of the alternative structural materials. The proposed form of the building, in relation to those for which the material is best suited, will normally have a major bearing on this. Any design decision has consequences and one of these is that the constraints and limitations of the material must thereafter be accepted. The adoption of masonry will normally result in the selection of a loadbearing-wall, panel-type structure (see Section 1.3.1). The advantages and disadvantages of masonry in relation to the other structural materials are summarised below.

5.2.2.2 Advantages

Strength
Masonry obviously has sufficient strength to perform as a structural material. It has moderate compressive strength but very low tensile strength and therefore has a limited capacity to resist bending. It performs best in situations in which compression predominates, as in the walls and piers of buildings of moderate height. It is also suitable for use in compressive form-active structures such as arches, domes and vaults.

Durability
Masonry is a durable material both physically and chemically. Except in the most exposed situations externally, and often even then, it can be left free of finishing materials. This both simplifies the detailing of buildings and provides a carcass which is virtually maintenance free.

Low cost
Due to the fact that relatively little energy is consumed in their production and due to their good availability at most locations, the basic cost of masonry units and the cost of transporting them to particular sites are low. Masonry structures can be built using simple, traditional techniques which do not require complicated plant or machinery. The construction process is therefore also relatively cheap; low cost is one of the principal advantages of masonry construction.

Appearance
Most masonry has a pleasing appearance which matures rather than deteriorates with age. A considerable variety of colours and surface textures is usually available and may be used to create a range of architectural effects.

Design flexibility
The fact that large structures can be assembled from small basic units allows complicated geometries to be achieved relatively easily with masonry and the material therefore offers considerable scope for imaginative design, subject always to the technical constraints (the

loadbearing wall, the arch, the vault and the dome). Another consequence of the method of construction is that other materials can be incorporated into masonry so as to augment its properties. The placing of steel reinforcement in the bedding planes to give masonry flexural strength is an example of this.

Fire resistance
Masonry performs well in fire; it is non-combustible and retains its structural properties at high temperatures.

Acoustic performance
Masonry walls form good acoustic barriers.

Thermal performance
Masonry walls provide a reasonable level of thermal insulation. Their high thermal mass is a further advantage which allows the creation of enclosures with good levels of passive environmental control.

5.2.2.3 Disadvantages

Lack of tensile strength
The lack of tensile and therefore of flexural strength is masonry's principal structural disadvantage and has restricted its use, in modern times, to loadbearing-wall-type structures. The selection of masonry therefore normally implies that the constraints of this type of structure be accepted in the planning of the building. The most obvious of these is the requirement that a plan consisting of parallel loadbearing walls be adopted and that, in multi-storey buildings, the plan be more-or-less the same at all levels.

The lack of tensile strength also makes difficult the construction of high-strength structural connections between masonry and other structural elements. This has tended to restrict the use of masonry in modern times to multi-cellular buildings in which none of the interior spaces is large.

Weight
In comparison to timber, masonry is relatively heavy and while this has advantages (it is responsible for the good performance of masonry as an acoustic barrier) it also has disadvantages. In particular, it results in high dead loads being imposed on supporting structures, such as foundations. It also affects the cost of transport to the site.

Porosity
Most masonry is porous and while this may not be a serious disadvantage so far as its use as a structural material is concerned, it does affect the structural design. In particular, where the material is used for external walls, it complicates the detailed design, which must be such as to prevent water from penetrating the building.

5.2.2.4 Conclusion
To sum up, the principal advantage of masonry is that it possesses a good combination of properties rather than that it performs outstandingly well with respect to any particular criterion. Its good appearance and durability, its reasonably high compressive strength, its low cost, its thermal and acoustic properties and its performance in respect of fire make it an ideal material for structural walls and this is, of course, its principal use. In modern practice masonry structures are usually based on post-and-beam arrangements and the particular properties of the material affect the forms which are adopted. The general principles which are followed in the planning of masonry buildings, and which take these into account, are outlined in Section 5.3.

The treatment of masonry which is given here is confined almost entirely to the use of brick and block manufactured components. These are by far the most commonly used constituents in the developed world, but masonry is also constructed in a variety of other materials such as natural stone and mud. Although these materials are not dealt with here specifically, the general principles of loadbearing masonry which are outlined in this chapter apply equally to construction in all brittle building materials.

5.3 The basic forms of masonry structures

5.3.1 Introduction

There are two basic types of masonry: plain masonry, in which bricks or blocks are simply bedded in mortar to form walls and piers, and reinforced masonry, which contains steel reinforcement bars in addition to the constituents of plain masonry (Fig. 5.36). Reinforcement bars provide tensile strength and produce a material whose structural properties are similar to those of reinforced concrete. Reinforced masonry tends to be used in conjunction with plain masonry in situations where walls are required to have flexural strength, such as where they are exposed to large out-of-plane loads. Its presence tends to have only a modifying effect on the forms of masonry structures, whose basic geometries have been determined by the properties of plain masonry.

In common with all buildings, loadbearing masonry structures must be capable of resisting two principal types of load: gravitational load, caused by the weight of the building and its contents, and horizontal load, caused by the action of the wind or of earthquakes. A number of form-determining factors result from this; the influences of each load type are considered separately.

5.3.2 Resistance to gravitational load

5.3.2.1 Introduction

The strategy for resisting gravitational loads in masonry structures is to use a post-and-beam arrangement in which the masonry forms the vertical elements and the horizontal elements are made from some other material, usually timber or reinforced concrete (Fig. 5.20). Two factors play a dominant role in determining the arrangement of walls which must be adopted. Firstly, the plan-form must be such as to provide support for all areas of floor and roof which have to be carried; this is the more important factor. Secondly, the walls must have adequate strength; they must not be over-stressed by the loads which they carry and neither must they become unstable by buckling. The second set of considerations tends to affect the thickness of walls rather than the plan-form of the building, but the need to provide lateral support to prevent wall buckling can affect the latter.

5.3.2.2 Provision of support for floors and roofs
The wall arrangements which are adopted to

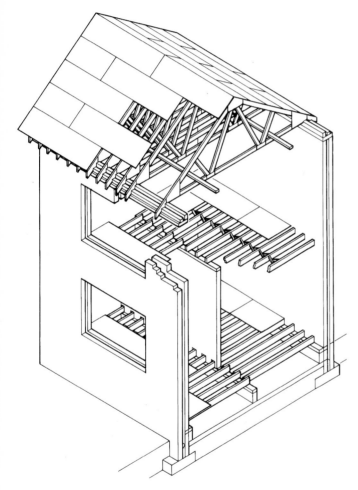

Fig. 5.20 Typical arrangement of structural elements in a loadbearing masonry building. The timber floor structures are carried on a parallel arrangement of loadbearing walls. The trussed-rafter roof, which is a more efficient type of structure than the rectangular cross-section floor joists and which carries less load, spans across the entire width of the building between the exterior walls. Additional non-loadbearing walls (not shown) are required in the across-building direction for stability.

provide adequate vertical support for all areas of floor and roof depend on the type of floor and roof system which is used and, in particular, on whether one-way- or two-way-spanning systems are employed. Typical one-way-spanning systems for masonry buildings are the traditional timber joisted floor and the traditional rafter or trussed-rafter roof (see Section 6.7.2); precast concrete slabs and one-way-spanning *in situ* reinforced concrete slabs are two other common examples (Fig. 5.21).

Examples of two-way-spanning systems are the *in situ* reinforced concrete floor slab and the steel space-deck roof structure (Fig. 5.22).

Four basic plan types are employed for masonry buildings: spine-wall, cross-wall, cellular and core-type (Fig. 5.23). Note that these are the basic forms in which only the primary loadbearing walls are shown. Extra structural walls are usually required for lateral stability, particularly in the spine- and cross-wall forms, and most building plans also contain a number of non-structural walls in addition to the structural walls, because these are necessary to achieve the required arrangement of internal spaces. The plans of real buildings, of which those shown in Fig. 5.24 are typical, are therefore much more complicated than the basic forms, which are shown

(a)

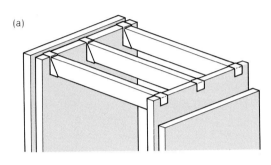

(b)

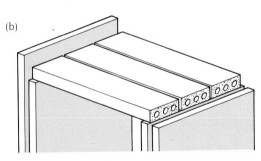

(c)

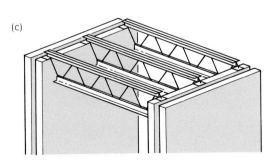

Fig. 5.21 Typical one-way-spanning floor structures in loadbearing-wall buildings.
(a) Timber joists supported on joist hangers.
(b) Precast concrete slabs.
(c) Lightweight steel joists.

(a)

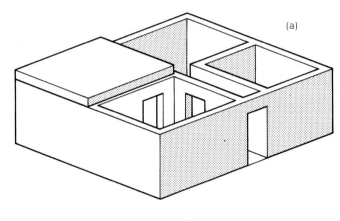

(b)

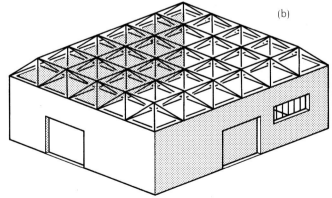

Fig. 5.22 Two-way-spanning structures.
(a) *In situ* reinforced concrete slab.
(b) Space framework.

165

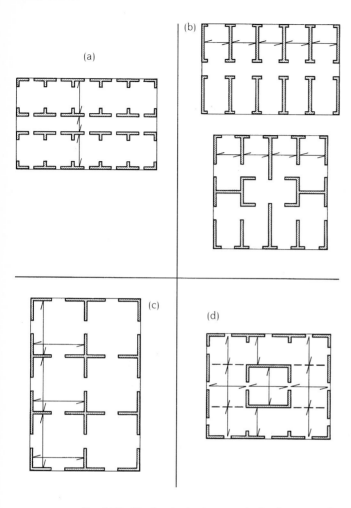

(a)

(b)

(c)

(d)

Fig. 5.23 The four basic plan types for loadbearing-wall structures. Only the loadbearing walls are shown.
(a) Spine-wall.
(b) Cross-wall.
(c) Cellular.
(d) Core-type.

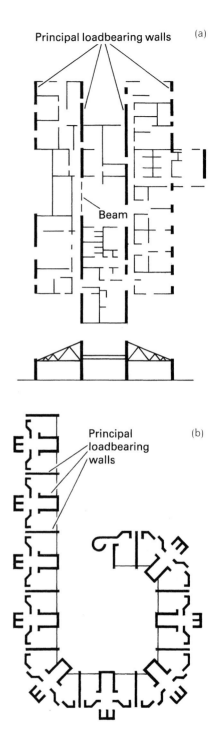

Principal loadbearing walls (a)

Beam

Principal loadbearing walls (b)

Fig. 5.24 Typical wall arrangements in actual buildings in which numerous non-structural partition walls are required for space-planning purposes.
(a) Spine-wall arrangement.
(b) Cross-wall arrangement.

diagrammatically in Fig. 5.23. All building plans must, however, contain some walls which conform to one of these basic plan-forms. Typical element dimensions for this type of building are given in Table 5.1.

Where a one-way-spanning floor system is used it is best supported on a parallel arrangement of loadbearing walls; the cross-wall and spine-wall arrangements are therefore particularly suitable for this type of construction. The spacing between the walls depends mainly on

Table 5.1 Data for elements in low-rise loadbearing masonry structures

Walls			single and 2 storey	> 2 storey
	External wall (cavity)	Inner leaf	102.5 mm	215 mm
		Outer leaf	102.5 mm	102.5 mm
	Internal wall	Loadbearing	102.5 mm	215 mm
		Non-loadbearing	102.5 mm	102.5 mm
Floors	Timber joist	Depth range 75 mm to 300 mm (see Table 6.10)		
	Reinforced concrete slab	Depth range 150 mm to 440 mm (see Table 4.1)		
Roof	Rafters and joists	Depth range 75 mm to 300 mm (see Table 6.11)		
	Trussed rafter	See Table 6.8		
	Truss	See Table 6.9		

the span capacity of the floor system. A span range of 3 to 5.5 m is normal for timber floors and 3.5 to 8 m for reinforced concrete. It is advantageous to keep all principal floor spans in a building (i.e. those of the main spaces) more-or-less the same by spacing the loadbearing walls equal distances apart, because this avoids the need for floor units which are of different depth and therefore simplifies the detailing of the building. It also maintains the stresses in the walls at a constant level and avoids the need to specify different strengths of masonry. Some spaces, such as corridors, in which spans are markedly different from the principal spans of a building are, of course, unavoidable. In buildings which have one-way-spanning floors and roofs, openings in walls are spanned by lintels; these are normally of precast concrete or steel.

The two-way-spanning floor system which is most commonly used in masonry buildings is the *in situ* reinforced concrete two-way-spanning slab, which is sometimes called a flat-slab. Individual slabs are supported by an arrangement of walls which is square on plan and this type of floor structure therefore works best in conjunction with cellular plan-forms in which each cell is approximately square. The high statical indeterminacy[6] of this form of structure allows it to be used with irregular plan-forms, however, and this is one of its

advantages over one-way systems. The high statical indeterminacy also means that the structural material is more efficiently used than in the one-way system, particularly if continuity is achieved over a number of approximately equal spans. This system is therefore capable of larger spans than are one-way systems and gives smaller slab depths for equivalent spans. The statical indeterminacy also makes it possible for walls to be designed with fairly large numbers of openings, without the need for local thickening of the slab or for lintels to be used to bridge the gaps.

It is desirable that the structural walls in multi-storey masonry buildings should be continuous throughout the height of the building and so the same basic plan-form must be adopted at every level. Where this is not feasible, as in certain building types in which larger interior spaces are required at ground floor level (for example, ground floor shops with houses above or an hotel with a restaurant and bar on the ground floor and bedrooms above) a special structural arrangement must be used at the level where the floor-plan changes. The most common solution to this problem is to use a reinforced concrete or steel frame at ground floor level; the reinforced concrete floor slab which forms the top of this is then used as a base for the loadbearing-wall structure above (Fig. 5.25). Alternatively the masonry itself can be reinforced by incorporating steel reinforcing bars into the bedding planes. This gives it

6 See Macdonald, *Structure and Architecture*, Appendix 3.

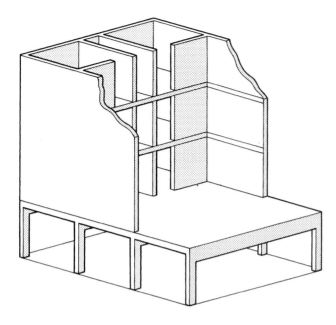

Fig. 5.25 Where large open-plan areas are required on the ground floor of a building in which the upper floors are multicellular, this can be achieved by the provision of a frame structure of steel or reinforced concrete to act as the base of a loadbearing-wall building.

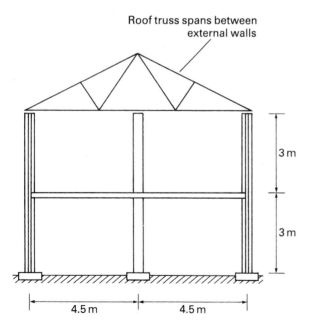

Roof truss spans between external walls

3 m

3 m

4.5 m   4.5 m

Fig. 5.26 Typical cross-section of a small spine-wall building in which the roof structure spans between the external walls.

similar properties to those of reinforced concrete and allows the walls themselves to act as beams spanning across the voids at ground level.

Where reinforced concrete floors are used they will normally be strong enough to support any non-loadbearing walls which are provided for space-planning purposes. There is therefore no structural requirement for non-loadbearing walls to be positioned vertically above one another and this makes possible minor variations in the plan between levels. Where timber floors are used, these will not normally be capable of supporting the weight of non-loadbearing masonry walls and it is therefore necessary that the plans at different levels should be more-or-less the same. Where minor variations occur, provision must be made to transfer the weight of non-continuous walls to the loadbearing walls on steel or reinforced concrete beams.

Roof structures are usually capable of longer spans than floor structures, especially if trusses or trussed rafters are used. In multi-storey structures the strategy which is adopted for supporting the roof is therefore frequently different from that for the floors in the same building. For example, in spine-wall construction the roof is often made to span between the external walls of the building and does not require support from the spine-wall (Fig. 5.26). The same system of supporting the roof on the external walls is frequently used when the floors span between cross-walls; the roof span is then at right angles to the span of the floors. Alternatively a purlin system can be used to support the roof as well as the floors on the cross-walls (Fig. 5.27). This relieves the external walls of any load-carrying function, which can be an advantage if large areas of glazing are required.

The thickness of the walls which are required in masonry structures depends on the individual circumstances of the building concerned and, in particular, on the load which each wall must carry. Any special requirements with regard to bond will also affect this. It is normal practice to specify cavity construction for the exterior walls, to prevent damp penetration,

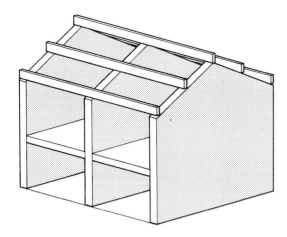

Fig. 5.27 Cross-wall arrangement with roof carried on purlins.

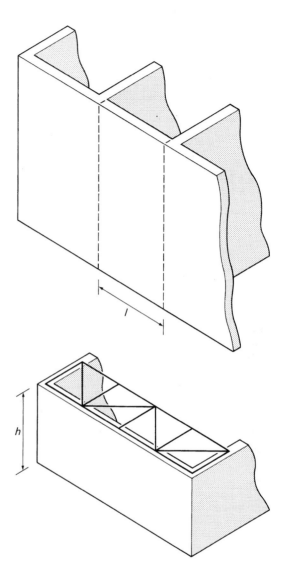

and for internal walls to be solid. In low-rise buildings of four storeys or less, the external walls will normally consist of two leaves which are one half-brick-length thick or its equivalent in blockwork (i.e. approximately 100 mm) separated by a 60 mm wide cavity to give a total wall thickness of approximately 260 mm. Normally only one of the leaves is loadbearing (usually the inner leaf), and frequently brick-work is used for the outer leaf and blockwork for the inner leaf. Interior walls can be one half-brick-length thick or its equivalent in blockwork unless, for reasons of appearance, a bond which requires headers is used. For high-rise buildings the inner, loadbearing leaf of external walls and the solid internal walls must usually be one brick-length thick or its equiva-lent in blockwork (i.e. approximately 200 mm).

5.3.2.3 Resistance to buckling

Masonry walls, being compression elements, are susceptible to buckling failure and the plans of masonry buildings have to be arranged so as to avoid the creation of walls which are excessively slender. The slenderness ratio of a wall is defined as the ratio of its effective length or its effective height to its width, whichever is the smaller (Fig. 5.28) and the British Standard for Loadbearing Masonry, BS 5628, requires that this should never be

Fig. 5.28 The slenderness of masonry walls, which is determined by the distance between locations at which the wall is provided with lateral support, must be limited to control buckling. Lateral support can be provided either by buttresses or orthogonal walls, in which case the slender-ness is determined by the horizontal distance between these, or by floor structures or horizontal-plane bracing girders, in which case the slenderness is determined by the vertical distance between these. In either case the measure of slenderness is simply the distance between the locations of lateral restraint divided by the thickness of the wall. The need to limit slenderness can sometimes affect the internal planning of buildings by requiring that buttresses or bracing walls be located in particular positions.

greater than 27. In practice, however, it is usually desirable to achieve a value which is considerably smaller than this (a maximum of around 20).

Given that the range of wall widths which is used in masonry is limited by the available dimensions of the masonry units, it is the effective lengths and effective heights of walls which must be adjusted by the designer of the building in order to control their slenderness. These are the distances between the horizontal and vertical lines along which the walls are restrained against lateral movement (movement out of their own plane (Fig. 5.28)). In the case of height the distance between lateral restraints will normally be the same as the storey height of the building because the roof and intermediate floor structures can normally be regarded as being capable of providing effective lateral restraint. The effective length of a wall is the distance between lateral support provided either by walls, bonded at right angles to the wall in question, or by substantial piers.

In most buildings, the storey height will be such that the slenderness ratio, calculated from the effective height, will be within the required limit. Where this is not the case it will not normally be possible to adjust the slenderness ratio by altering the storey height because many factors besides those concerned with structural performance influence the choice of storey height. In these cases the required slenderness ratio must be achieved by altering the plan so as to reduce the effective length of the wall, either by inserting buttresses or by altering the positions of walls which are perpendicular, on plan, to the wall in question.

It is important that the structural elements which are considered to provide lateral support for a wall are in fact capable of doing so. Two factors affect this, the strength of the supporting elements and the way in which they are attached to the wall. Buttressing walls and concrete floors and roofs will normally have adequate strength. The ability of timber floors or roofs which are sheathed with boarding to act as horizontal diaphragms may be in doubt, however, but where this occurs, they can be

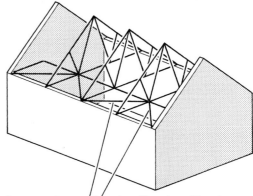

Extra members are introduced into the ceiling plane to form a horizontal-plane girder which provides lateral support for the top of the wall

Fig. 5.29 A bracing girder is located at eaves level and acts in conjunction with the gable walls to provide lateral support for the top of the side wall.

strengthened either by sheathing in structural plywood, so as to form a strong horizontal box beam, or by incorporating into the floor or roof structure a system of diagonal bracing (Fig. 5.29).

5.3.2.4 Summary of the form-determining factors which are derived from the need to resist gravitational load
The need to resist gravitational load exerts two major influences on the plan-forms which are adopted for loadbearing masonry buildings. Firstly, the requirement to provide support for all areas of floor and roof favours the use of certain plan geometries; secondly, the need to prevent buckling requires that the slenderness ratios of walls should not be unduly large. To satisfy these requirements the following general rules are observed when planning masonry buildings.

1 The arrangement of supporting walls is tailored to the type of roof or floor system which they will support. For one-way-spanning systems parallel sets of walls are used and placed at approximately equal distances apart. For two-way-spanning systems a cellular arrangement is adopted in

which the individual cells are approximately square and approximately the same size.

2 The plan geometries of multi-storey buildings are kept more-or-less the same at every level. It is particularly important that loadbearing walls should be continuous throughout the entire height of a building.

3 Attention is paid to the slenderness ratios of walls and columns. Storey heights are no larger than is necessary and long sections of plane wall are avoided. A wall thickness is selected which gives a slenderness ratio of less than 20.

5.3.3 Provision of overall stability and lateral strength

5.3.3.1 Introduction
As with most types of structure, the issue of the overall stability of a masonry building can be considered together with that of its ability to resist horizontal load. If, in addition to gravitational load, the building can resist horizontal load from two orthogonal directions then it will also be geometrically stable.

The action of horizontal load on a masonry building has two consequences which affect its planning. Firstly, it requires that the building, taken as a whole, should have the ability to resist its effect (Fig. 5.30a). The building must,

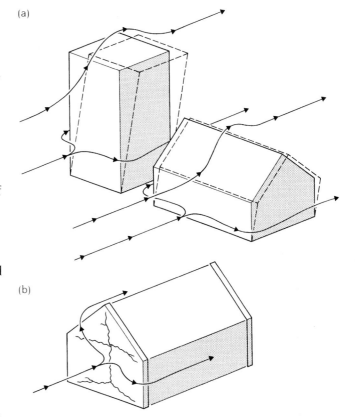

(a)

(b)

Fig. 5.30 The effect of wind loading on a masonry structure.
(a) The structure, taken as a whole, must be capable of withstanding the horizontal load.
(b) Individual walls must have sufficient bending strength to resist the out-of-plane pressure loads.

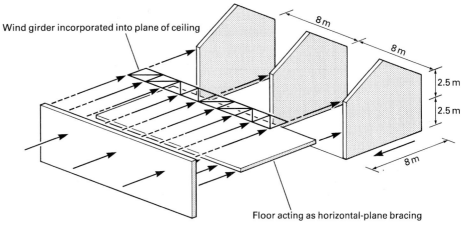

Fig. 5.31 Typical wind-bracing system for a loadbearing-wall structure. Wind pressure loads are applied initially to the surfaces of the external walls and are transmitted by the horizontal-plane floors and wind girders to walls which are parallel to the wind direction and which act as diaphragm bracing in the vertical plane.

Wind girder incorporated into plane of ceiling

8 m

8 m

2.5 m

2.5 m

8 m

Floor acting as horizontal-plane bracing

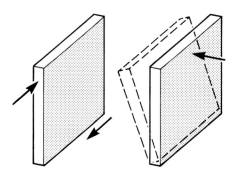

Fig. 5.32 Individual plane walls are ineffective at resisting out-of-plane loads but can resist in-plane loads.

in other words, contain a structural system which is capable of conducting the lateral loads from the points at which they are applied, to the foundations, where they can ultimately be resisted (Fig. 5.31). This property will also guarantee stability, as noted above. Secondly, the horizontal load subjects the external walls to out-of-plane forces which they are ill equipped to resist, due to their low strength in bending (Fig. 5.30b). Provision must therefore be made either to give them the necessary strength to resist these loads or to provide them with lateral support.

5.3.3.2 The lateral strength of the building, taken as a whole
The strategy which is adopted to provide a masonry building with lateral strength, and therefore also stability, is usually based on the very good performance of walls in resistance to in-plane forces (Fig. 5.32). Buildings are therefore planned so that wind forces are conducted to the foundations by walls which are more-or-less parallel to their direction.[7] This means that walls must be provided in two orthogonal directions on plan. In practice, it is rarely difficult to achieve this because the space-

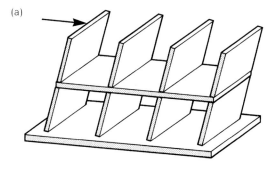
(a)

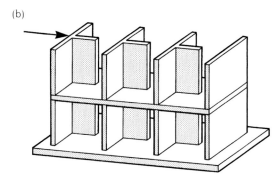
(b)

Fig. 5.33 This cross-wall arrangement required non-loadbearing bracing walls for stability.

planning requirements of masonry buildings will normally produce a plan which contains walls which run in the two principal directions, even though the primary loadbearing walls may be parallel to each other in only one of these. The plans in Fig. 5.24, for example, are satisfactory in this respect. Figure 5.33a shows an arrangement which is not satisfactory because there is no wall which can resist wind load as an in-plane force in one of the building's principal directions. Additional walls must therefore be added to render the building stable (Fig. 5.33b).

The horizontal parts of masonry structures play an important part in the resistance of horizontal load and they can act in two capacities. For a particular direction of horizontal load they provide a direct tensile or compressive link between bracing walls and other walls which require to be stabilised. They can also act as horizontal-plane bracing (Fig. 5.31). Normally the horizontal structure in a masonry

---

7 The building must obviously be capable of resisting wind load from any direction. This condition is satisfied if resistance is provided in two orthogonal directions.

building will be required to carry out both of these functions and it must be designed accordingly, and suitably attached to all of the walls. Reinforced concrete floors provide the best horizontal structural elements for these purposes. Where timber elements are used, their ability to act collectively as a horizontal diaphragm must always be considered. In cases where they cannot do so, a horizontal bracing girder must be incorporated into the plane of a roof or floor to give it the ability to act as horizontal-plane bracing (Fig. 5.31).

In low-rise buildings of up to four storeys the magnitudes of the lateral loads due to wind are not high and the resistance of these is rarely a critical factor in the design. The building will be satisfactory so long as it contains some walls in two orthogonal directions and any single-plane walls which are present are connected, by an adequate horizontal system, to other wall groups which are capable of bracing them. The exact disposition of the loadbearing and bracing walls is not critical although a symmetrical arrangement of bracing walls will produce the most satisfactory structure for the resistance of lateral load.

In tall multi-storey buildings of five storeys or more, the resistance of wind load can, however, be a critical consideration. The additional factors which affect their planning are reviewed in Section 5.3.4.

5.3.3.3 Resistance of individual walls to out-of-plane forces
Wind pressure is an out-of-plane load on the external walls of buildings. It produces bending-type internal forces in walls and can result in the generation of levels of tensile stress which the masonry cannot resist (Fig. 5.30b). The task for the designer is to contain the tensile stress which develops. This can be done by adopting a structural form which either limits bending action or which neutralises, using axial compressive stress, any tensile bending stress which does develop. Bending action is countered by minimising span. The external walls of masonry buildings span either vertically, between floors, or horizontally, between buttressing walls (Fig. 5.34). If the

(a)

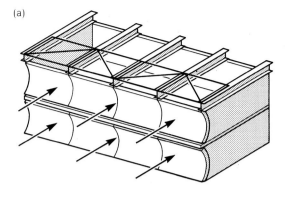

(b)

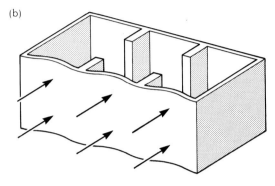

Fig. 5.34 In response to out-of-plane wind loads, the external walls of loadbearing wall buildings span either vertically between horizontal-plane bracing elements or horizontally between buttressing walls.

strategy being adopted to limit bending action is that of providing buttressing walls, then these have to be spaced closely. This obviously affects the planning of the building.

If the strategy for limiting the bending stress caused by out-of-plane wind pressure on external walls is one of neutralising the tensile stress with compressive stress, the walls must be arranged to span vertically between floors in response to wind loading. This ensures that any tensile stress acts on the horizontal cross-sectional planes where it can be neutralised by the compressive stress produced by the gravitational loads. The magnitude of the tensile bending stress can be minimised by ensuring

---

8 See Macdonald, *Structure and Architecture* Appendix 2 for an explanation of this term.

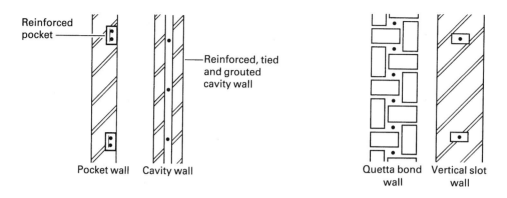

Fig. 5.35 Plans of walls showing various arrangements for incorporating vertical reinforcing bars. These give the walls the ability to span vertically between floor and roof structures in response to out-of-plane loads.

that the second moment of area[8] of the wall cross-section is maintained high, by making the wall reasonably thick.

Masonry, as mentioned above, can be given the bending strength which is necessary to resist high out-of-plane load by reinforcing it with steel bars. Its behaviour is then similar to that of reinforced concrete. The technique is usually used to augment the strength of walls which must carry very large bending loads, such as retaining walls or the external walls of high buildings. Vertical bars provide the bending strength to span vertically in response to out-of-plane forces. These must either be incorporated into the voids of special bricks or blocks, or accommodated in vertical channels provided by special bonds (Fig. 5.35). Walls can also be given the strength to span vertically by incorporating reinforced concrete or steel columns into them. Horizontal bars in the bedding joints of masonry or within special blocks (Fig. 5.36) give walls the strength to span horizontally. These techniques do not affect the overall form which must be adopted for a building, except insofar as they permit large areas of plane wall to be adopted.

5.3.3.4 Summary of form determining factors which result from the need to resist horizontal load
The principal features which are incorporated into the geometries of masonry buildings in order that they can resist horizontal load effectively are as follows.

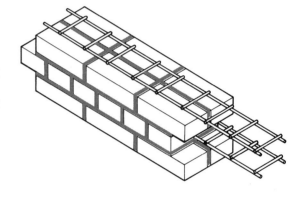

Fig. 5.36 Incorporation of reinforcement into the bedding planes gives masonry the ability to span horizontally between buttressing walls in response to out-of-plane loads.

1 The plan must contain walls orientated in two orthogonal directions.

2 The horizontal parts of the structure must provide an effective structural link between the walls. They must therefore have adequate strength and must be adequately fixed to the walls.

3 The walls are disposed in as symmetrical an arrangement as possible on plan. This is not essential in low-rise buildings, where wind loads are small, but becomes increasingly important as building heights increase.

(a)

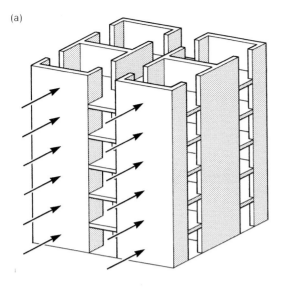

(b)

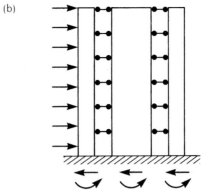

Fig. 5.37 Tall multi-storey loadbearing-wall structures behave as 'bundles' of cantilevers in response to wind loading. The structural action is most effective when individual wall groups are as large as possible.

4 The bending action which results from the application of out-of-plane forces to the external walls is minimised by providing lateral support at frequent intervals and thus lowering the span. Attention is also given to the selection of the cross-sectional dimensions of external walls to ensure that tensile bending stresses in response to out-of-plane loads do not become excessive.

5.3.4 Tall multi-storey buildings

Tall multi-storey buildings may be regarded as behaving like 'bundles' of vertical cantilevers when subjected to wind loading (Fig. 5.37). These perform best if the walls are arranged into groupings which are as large as possible, and which are bonded together at their vertical junctions (Fig. 5.38). Symmetry in the disposition of walls is also desirable, to prevent twisting of the building under the action of asymmetrically applied horizontal load.

5.3.5 Tall single-storey buildings

Tall single-storey buildings are buildings which consist predominantly of a single large interior space (sports hall, swimming pool). If these are constructed in loadbearing masonry the provision of walls which can withstand gravitational load is rarely a problem. A post-and-beam form can be adopted in which masonry walls support

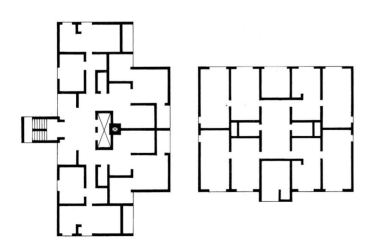

Fig. 5.38 Typical planforms for tall multi-storey loadbearing-wall structures. Each contains several very large wall groupings which have complex plans. These will make the largest contribution to the resistance of wind loading.

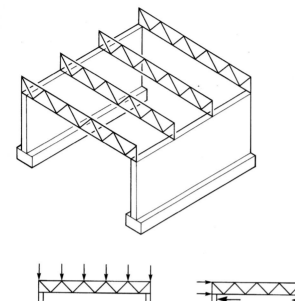

Table 5.2 Spans and dimensions for roof elements in tall single-storey masonry structures

Beam Span (m)	Beam depth (mm)			
	Universal beam	Castellated beam	Parallel chord truss (hot-rolled elements)	Laminated timber beam
10	450	500	1000	700
15	600	700	1250	900
20	700	800	1500	1200
25	800	900	1750	–
30	900	1200	2000	–
35	–	1300	2250	–
40	–	1300	2500	–

Fig. 5.39 Tall single-storey building. A roof structure of timber or steel girders is supported on high masonry walls. Such buildings are the modern equivalent of the vaulted hall (see Section 5.2.1).

a roof structure of steel or timber-truss primary elements (Fig. 5.39). Typical depths for various types of roof structure are given in Table 5.2.

The strategy which is normally adopted to resist horizontal load is fairly standard. Vertical-plane bracing is provided by the walls which are parallel to the direction of the lateral load and these are linked to all other parts of the building through horizontal-plane bracing (Fig. 5.40). In the case of these single-storey buildings the horizontal-plane bracing must normally be provided by incorporating a wind-girder into the plane of the roof.

External walls which are exposed to out-of-plane wind pressure span vertically between the wind girder and the foundations and, as the walls are high, the resulting bending action can be large (Fig. 5.40). Because the compressive stress due to gravitational load in the roof will be only moderate in a single-storey building, it is necessary that the tensile bending stress should not be high. The latter is controlled by the expedient of increasing the second moment of area of the wall cross-section, and this is a factor which can have a major effect on the planning of the building.

If the wall is relatively low (say around 3 m), normal cavity construction, with perhaps a thicker than normal loadbearing leaf or with pier stiffening of the loadbearing leaf (Fig. 5.41a), is usually sufficient to prevent the bending stress from being excessive, but if it is higher, then relatively sophisticated systems must be used. In the range 4 m to 5 m one of the most economical forms of construction is the fin-wall (Fig. 5.41b) in which one of the leaves is stiffened with very large piers. The wall then behaves as a series of T-beams in response to lateral load. Fins are normally one-half to two brick-lengths wide and four to eight brick-lengths deep and are spaced 3 m to 5 m apart. For wall heights in the range 5 m to 10 m a form of construction called diaphragm walling is necessary (Fig. 5.41c). This involves separating the leaves of the wall by a gap which is much larger than is used in normal cavity construction, that is by one or two brick-lengths, and providing a shear connection between the leaves in the form of diaphragms at closely spaced intervals (1 m to 1.5 m). The wall then acts as a box-beam in response to

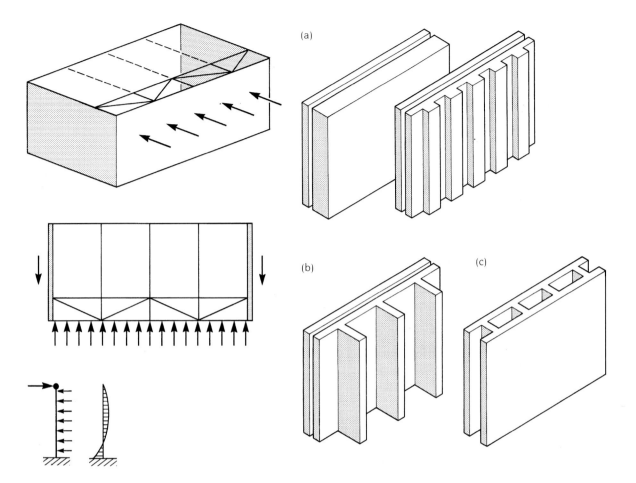

Fig. 5.40 Action of tall single-storey building in response to wind loading. The high walls span vertically between the foundation and the wind girder at roof level, which transmits the load to the end walls. The side walls here act as propped cantilevers and can be subjected to high levels of bending moment.

Fig. 5.41 Various strategies can be adopted to give the walls of tall single-storey buildings sufficient bending strength to resist out-of-plane load due to wind.
(a) Buttressed cavity wall (suitable for heights up to 3 m).
(b) Fin wall (suitable for heights up to 5 m).
(c) Diaphragm wall (suitable for heights up to 10 m).

bending loads. Typical dimensions for all of these types of wall are given in Table 5.3.

5.3.6 Provision for accidental damage
There are certain types of very high intensity load whose occurrence is so rare that it is not economic to make walls strong enough to resist them. Examples are the collision of a heavy vehicle with a building or an explosion which results from the ignition of leaking gas. Because buildings are not sufficiently strong to resist these, damage or partial collapse results

if they occur, but the risk of this happening is taken to be sufficiently low to be acceptable. It is necessary, however, that the stability of the whole building should not be endangered if a single structural element, such as a wall or an area of floor, should collapse due to the occurrence of a freak load. In other words, a progressive collapse should not result from an incident in which a part of the building only is directly involved.

The UK Building Regulations require that structures should be designed so that progres-

Table 5.3 Wall dimensions for tall single-storey masonry structures

Wall height	Plain cavity wall	Fin wall		Diaphragm wall	
(m)	t_i (mm)	t_p (mm)	T (mm)	D (mm)	S (mm)
3	215	–	–	–	–
4	327.5	665	440	–	–
5	440	890	440	–	–
6	–	1012.5	440	557.5	1360
7	–	1340	440	557.5	1135
8	–	1340	440	557.5	910
9	–	–	–	665	1135
10	–	–	–	665	910
	Fig. 5.42a	Fig. 5.42b		Fig. 5.42c	

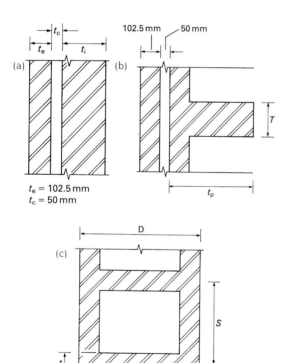

Fig. 5.42 Plan dimensions of walls.
(a) Plain cavity wall.
(b) Fin wall.
(c) Diaphragm wall.

sive collapse cannot occur and specify limits within which the collapse which is caused by a single incident must be contained. These regulations apply to all types of building but are particularly relevant to the overall planning of masonry structures which are more vulnerable to this form of collapse than other types. As with all of the Building Regulations, however, the requirement is no more difficult to meet than that which a responsible designer would in any case wish to impose.

In practice the satisfaction of the requirements in respect of progressive collapse is not difficult with a masonry structure. Both common sense and the Building Regulations dictate that, in these exceptional circumstances, the concern should be with the preservation of life rather than with the preservation of the building itself. A considerable amount of damage to the building is therefore acceptable, following an incident, so long as the occupants are neither injured nor trapped as a consequence of it. The factor of safety against the total collapse of a damaged building can be very low, and a figure of 1.05 is specified in the Building Regulations.

Most masonry structures are capable of 'bridging' a gap in the structure with this degree of safety if certain rudimentary precautions are taken during the design. These usually amount to the incorporation of certain details into the wall-to-wall and wall-to-floor junctions so as to improve the integrity of the structure, and the insertion of a certain amount of reinforcement into key walls and piers. It is not usually necessary for any major alteration to be made for this purpose to the structural form which is required to resist the normal gravitational and horizontal loads. For this reason the specific design requirements in respect of the prevention of progressive collapse are not dealt with here.

Chapter 6

Timber structures

6.1 Introduction

Timber is a structural material with a useful combination of physical properties. Although its strength is not high (typical design stress values are in the range 5 to 20 N/mm^2 compared to equivalent values for steel of 150 to 250 N/mm^2), timber is, like steel, more-or-less equally strong in tension and compression. It can therefore withstand bending and can be used to make every kind of structural element. Due to the origin and nature of the material, timber is available normally in the form of slender, linear elements and this favours its use in framework arrangements.

The ratio of strength to weight of timber is high, and is comparable with that of steel, with the result that, although, for a given size of cross-section, timber elements are not so strong as those of steel, they are much lighter. Timber is therefore a lightweight material, capable of providing structural elements which are of low dead weight, but which are nevertheless reasonably strong and tough.

One of the problems associated with the structural use of timber is that due to its arboreal origins individual elements are relatively small. The construction of large

Fig. 6.1 Timber houses with skeleton-frame structures in which the principal elements are A-frames of plywood box-section. A characteristic of timber skeleton frameworks is that the primary structural elements are normally fairly large and must be incorporated into the aesthetic scheme of the building [Photo: Finnish Birch Plywood Association].

structures therefore involves the joining together of many separate components. Joints and connections are therefore an important aspect of the technology of structural timber.

6.2 Timber and architecture

6.2.1 Introduction

Of the four principal structural materials, timber is one which is not directly associated with a major architectural style or movement in the Western architectural tradition, although significant timber building traditions (for example, the 'stick' and 'shingle' styles of North America) have occurred. Other architectural traditions, such as those of China or Japan, have, however, produced significant timber styles.

Although no major Western architectural style is associated with timber, the contribution of the material to the development of Western architecture has nevertheless been considerable. Its principal structural use has been as the horizontally-spanning elements in post-and-beam structures in which the vertical elements were of masonry. As in the case of structures constructed entirely of masonry, it will be necessary here to consider historic examples in order to review the full contribution which timber has made to architectural expression.

Before considering the architectural qualities of timber it is necessary to say something of the range of structural arrangements for which it is suitable. Although there is considerable variation in the forms and layouts which have been created, almost all timber floor and roof structures conform to a similar basic layout. The roof or floor covering of boards, slates, tiles, etc. is normally supported on a series of closely spaced parallel elements. The timber beam floor (Fig. 6.2) is perhaps the simplest example of this. The modern trussed-rafter roof is a more sophisticated version of it (Fig. 6.27). The common characteristic of these arrangements is that the individual elements, be they beams or trussed rafters, carry relatively small areas of floor or roof and are therefore lightly loaded. They span one way between parallel supports which may be loadbearing walls or may be primary beams in a skeleton frame.

The spacing of the structural elements in this type of arrangement is close and is determined principally by the span capability of the material which the timber elements support (floor or roof boarding) but also by the need to maintain the load carried by each timber element at a suitably low level.

Where large spans are involved and the sizes of the main elements are therefore large it becomes uneconomic for the spacing of these to be close. For large spans, therefore, a hierarchical arrangement of primary, secondary and even tertiary elements provides a better solution. An example of this is the purlin roof (Figs 6.44 and 6.45). In the arrangement shown in Fig. 6.44 the primary elements, which span

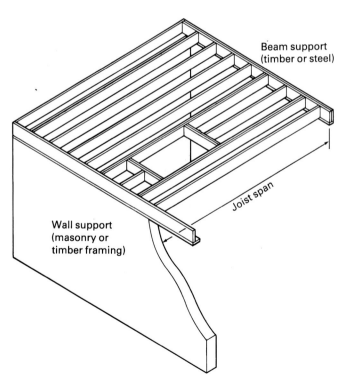

Beam support (timber or steel)

Joist span

Wall support (masonry or timber framing)

Fig. 6.2 In this typical timber floor structure the elements are at close spacings and each carries a relatively small amount of load.

Fig. 6.3 Primary–secondary system. The primary elements here are laminated timber portal frames. These carry purlins which in turn support the roof and wall cladding.

the full width of the space, are trusses. Purlins (secondary elements) span between these and support rafters, which are tertiary elements. A characteristic of the hierarchical arrangement of elements is that the primary elements carry the greatest amount of load and must normally therefore be relatively complex built-up arrangements such as triangulated trusses. The secondary and tertiary elements are closely spaced and carry relatively small amounts of load across small spans, as in the case of the purlins in Fig. 6.3 and the rafters in Fig. 6.44.

The elements of which timber structures are composed may take a number of forms (Fig. 6.4). The simplest of these is the solid beam. This carries load by pure bending action and is the least efficient type of structural element.[1] The maximum span which is possible is determined by the largest size of cross-section which is available. In the present day the maximum span possible with timber floor joists is around 6 m unless larger than normal sizes of cross-section, which are now difficult to obtain and therefore expensive, are used. In Greek and Roman antiquity, when larger sizes were apparently readily available, beam spans of up to 12 m were used.

The rafter arrangement (Fig. 6.42) is more efficient than that of the beam because the elements carry a combination of axial internal force and bending. This, together with the fact that the individual elements do not cross the entire span, allows a greater span to be achieved but has the disadvantage of imposing horizontal thrust at the points of support. The rafter system therefore carries with it the same problem as the masonry vault, which is the generation of horizontal forces at the tops of supporting walls. The magnitudes of these can be reduced if a tie is introduced (Fig. 6.43) and if the tie is placed at the level of the supports the lateral thrust is eliminated. The tie need not be a single piece of timber but can be jointed. It can also be supported from the apex of the rafters by a second, vertical tie (Fig. 6.43).

The tied rafter with the tie at support level, and secondary vertical tie, is in fact a simple version of a fully triangulated truss or trussed rafter (Figs 6.27 and 6.28), which is more efficient than any of the various rafter systems because the majority of the constituent sub-elements carry axial internal forces rather than bending.

The fully triangulated truss is a type of element which is frequently used for long spans in the present day. The advantages of triangulation seem not to have been fully understood until quite late in the history of Western architecture, however, and it was not until the nineteenth century that widespread use was made of fully triangulated, highly efficient truss systems, which exert no horizontal loads on

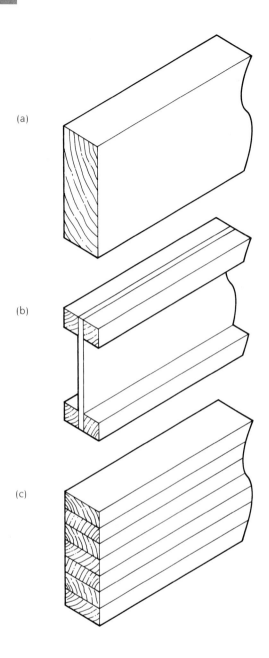

(a)

(b)

(c)

Fig. 6.4 A selection of timber element types.
(a) Sawn-timber joist – the simplest type of element which is produced in a wide range of sizes.
(b) Plyweb beam. The web of this built-up-beam is of plywood and the flange elements of sawn timber. Rectangular box sections are another common type of plyweb beam.
(c) Laminated timber beam. Large cross-sections can be built up by the laminating process which allows curved arch and portal frame elements to be produced as well as straight or tapered beams.

---

1   See Macdonald, *Structure and Architecture*, Chapter 4.

supporting walls. Most earlier truss systems were hybrid arrangements in which some triangulation was used but which also relied on rigid joints or continuity of elements through joints for satisfactory structural performance (see Fig. 6.29). Both of these expedients induce bending into the sub-elements of the truss and reduce their efficiency. The consequence was that larger cross-sections had to be used than would have been necessary had full triangulation been adopted.

6.2.2 The architectural significance of the use of timber as the horizontal elements in large-scale loadbearing-wall masonry structures

Timber beams were used in the roofs of the temples of Greek antiquity – probably in the form of simple beam systems (Fig. 6.5). It seems unlikely that their presence affected the external appearance of the buildings, which is the chief legacy of Greek builders to the vocabulary of Western architecture. The use of timber would have allowed larger interior spaces to be created than would have been possible had the buildings been constructed entirely in stone but, as the Greeks seem not to have used any form of built-up-beam or truss, the maximum span (around 12 m) was determined by the largest size of timber which was available. There is some evidence that the Greeks were aware of the principle of trussing but that, as in the case of the arch, they chose not to use it.[2]

The Romans, who were more adventurous technically than the Greeks, did adopt the principle of trussing. A basic triangulated arrangement was described by Vitruvius and he also provided a description of a timber-roofed basilica with a span of 60 Roman feet (approx. 20 m). A further example from the first century BCE is that of the basilica at Pompeii. The conjectured reconstruction of this building by Lange[3] (Fig. 6.6) shows a truss-and-purlin arrangement.

Examples of later Roman timber-roofed basilicas are that of the original St Peter's in Rome (Fig. 6.7) and the church of St Paul's Outside the Walls, also in Rome (Figs 6.8 and

Fig. 6.5 Cross-section of the Parthenon, Athens, 5th century BCE. In this conjectured reconstruction of the Parthenon the roof structure consists of timber beam elements (shown shaded). The spans are kept short by the subdivision of the cross-section by walls and rows of columns. Note the very deep beam which is required to cross the central part of the interior, between the rows of columns which flank the statue, and which carries the posts which support the ridge beam (after Coulton).

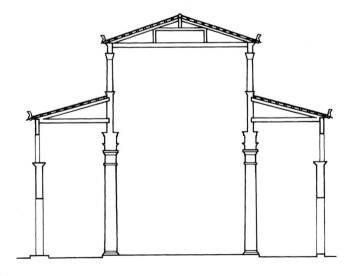

Fig. 6.6 Cross-section of Basilica at Pompeii, 1st century BCE. A form of semi-trussed timber structure is used here to span the wide central space (after Lange).

2 See Mark (ed.), *Architectural Technology up to the Scientific Revolution*, Cambridge, MA, 1993, Chapter 5.
3 Lange, *Basilica at Pompeii*, Leipzig, 1885.

Fig. 6.7 St Peter's Basilica, Rome, 4th century CE. A fully triangulated timber structure, which exerts no lateral force on the walls, spans the central part of the building. The use of this type of timber structure has allowed the creation of a large interior space without the need for the elaborately buttressed supporting walls which would have been required had a masonry vault been used (after Mark).

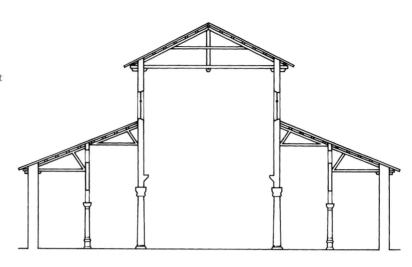

Fig. 6.8 *Left.* St Paul's Outside the Walls, Rome, 4th century CE. This perspective drawing illustrates well the advantage of using a triangulated timber roof structure rather than a vault. The thinness of the walls is well shown (after Piranesi).

Fig. 6.9 *Below.* St Paul's Outside the Walls, Rome, 4th century CE. Cross-section (after Rondelet).

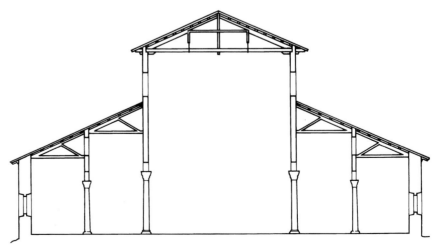

6.9). Both of these buildings are of the fourth century CE. They each consisted of a high central nave flanked by lower aisles. The naves, which were each approximately 24 m wide, were roofed in timber by truss-and-purlin systems and the flanking aisles by lean-to structures, also of timber. The use of tied trusses over the naves eliminated side thrust at wallhead level and enabled the builders to adopt very slender walls. The buildings were of light and delicate appearance, as is well illustrated by the seventeenth-century drawing of St Paul's Outside the Walls by Piranesi (Fig. 6.8).

The importance of the timber-roofed basilica was that it was a building with a relatively large interior which did not have the massive buttressing walls which were required to support a masonry vault (the only other way of achieving a large span at the time). It was an early example of a building type which would make a very significant contribution to the development of Western architecture – the large building with large interior spaces and *thin* walls. It was the development of the technology of the timber truss, from the Roman period onward, which made this possible.

Examples of this type of building can be found in all subsequent periods of western architecture. The palaces and large houses of the Italian Renaissance, the country houses and public buildings of the classical period in Britain, Northern Europe and America (Fig. 6.10) and churches from all subsequent periods have structural armatures of this type. In most of these buildings the timber structures which spanned the large interior spaces were entirely hidden from view and made no obvious contribution to the architecture. The forms of the buildings would have been impossible to achieve, however, without the technology of the timber truss.

In the nineteenth century timber gave way to iron and then to steel as the principal material from which large trusses for long-span roofs were constructed. The tradition of using large timber trusses did not die out, however. Twentieth-century examples have frequently

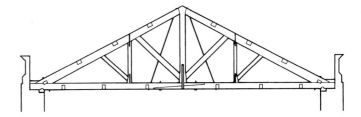

Fig. 6.10 Banqueting House, London, 1619–22. Inigo Jones, architect. The roof trusses which span the width of the Banqueting House in London (17 m approx.) are typical of the structural arrangements which were used to create large interior spaces in the buildings of the European Renaissance.

been exposed in the interior of buildings and used as part of the architectural language.

6.2.3 Timber loadbearing-wall structures

A timber loadbearing-wall structure is a post-and-beam arrangement in which both the vertical and the horizontal structural elements are of timber. The horizontal structures are similar to those which are used in conjunction with masonry and are beamed floors or rafter or trussed roofs. In timber loadbearing-wall structures the vertical elements are also of timber and consist of loadbearing walls formed by closely spaced timber sub-elements tied horizontally at regular intervals by further timber sub-elements. The building type has existed from the beginning of the Western architectural tradition and has taken two distinct forms – the half-timbered building and the timber wallframe building.

The origins of the half-timbered building lie in prehistory but the building type did not become significant architecturally until the eleventh century. It remained so until the seventeenth century. In the half-timbered building the walls consist of large timber elements, frequently roughly hewn and with cross-sections which were more-or-less square, tied at each storey level by horizontal elements of similar dimensions (Fig. 6.11). A third set of elements, inclined to the vertical at between 45° and 60°, was provided to brace the structure. The joints between the elements (Fig. 6.12) were of the type which required the

Fig. 6.11 The medieval half-timbered house was basically a loadbearing-wall structure. The walls consisted of vertical elements at relatively close spacing tied by horizontal elements at each storey level. The structure is stabilised by the massive masonry chimney and by the inclined, often curved, corner elements. The sizes of the cross-sections of the main element were large.

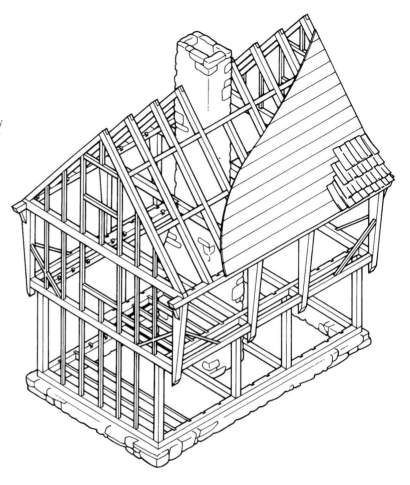

cutting away of material (mortice-and-tenon, halved, dovetail) and the resulting inefficiency was one of the reasons why timbers of large cross-section were used. The loadbearing walls were arranged parallel to each other and connected by horizontal beamed floors and rafter or trussed-roof structures. The building was completed by non-structural elements. The wall infilling was entirely non-structural and was made from a variety of materials. Brickwork or plaster on a woven mesh of thin timbers were commonly used. Normally, the structural timber was exposed on the exterior of the buildings, but occasionally the walls were rendered with plaster or lime.

The European tradition of the half-timbered building occurred north of the Alps and was associated with the rise of the mercantile middle class which occurred after the Reformation. The building type was used extensively for merchants' houses and public buildings in the trading towns of northern Europe.

The form of the buildings was predominantly rectilinear and the plan was that of the loadbearing-wall structure – parallel arrangements of walls supporting one-way-spanning floors and pitched roofs. Often, the buildings were of considerable height (five or six storeys) and the patterns formed by the exposed structural timbers on the external walls were frequently used as ornamentation and regularised and formalised for aesthetic purposes (Fig. 6.13).

The construction of half-timbered buildings declined in the seventeenth century for economic reasons. The very large sections of timber which were required became increasingly scarce, and therefore expensive, and the labour-intensive construction process, with its requirement for complicated joints, became uneconomical in the context of a growing materialistic, industrialised society in which labour costs were destined inexorably to rise. Thus did a building system which had considerable architectural potential pass out of existence. Its revival, in a modern or post-modern context, should not be ruled out, in view of its potential as a constructional system for a sustainable architecture.

The same economic forces which caused the demise of the half-timbered building led eventually to the development of a new type of timber loadbearing-wall structure, the timber wallframe. The wallframe type of building, which was developed in the nineteenth century in North America, can be seen as an inevitable consequence of the increasing use of manufactured products in building (Fig. 6.14).

The wallframe building consisted of a series of timber loadbearing walls carrying timber floor and roof structures in a conventional rectilinear layout not dissimilar to that of the half-timbered building. The loadbearing walls were formed by closely spaced timber elements tied horizontally at the levels of the roof and floors. In the wallframe building, however, the timber elements were produced

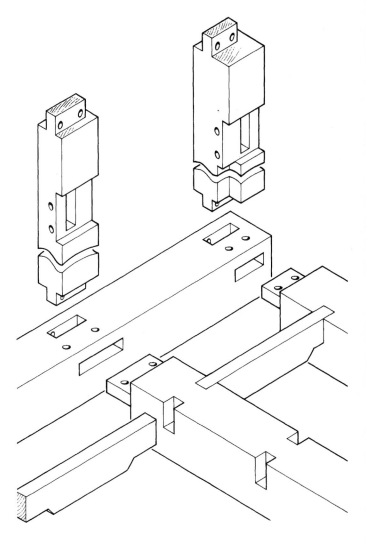

Fig. 6.12 Joints in half-timbered buildings were of the traditional carpentry type which involved the removal of large areas of cross-section. This is one of the reasons why such large overall cross-sections were required.

Fig. 6.13 The exposed structure contributes to the aesthetic appeal of this modest example of a half-timbered building.

Fig. 6.14 The timber wallframe is a loadbearing-wall structure built up from sawn-timbered elements. It is the industrialised equivalent of the half-timber building. The fastening elements are nails which allow the joints to be made without removing significant parts of the cross-sections. The sizes of the cross-sections are therefore considerably smaller than those used in half-timbered buildings.

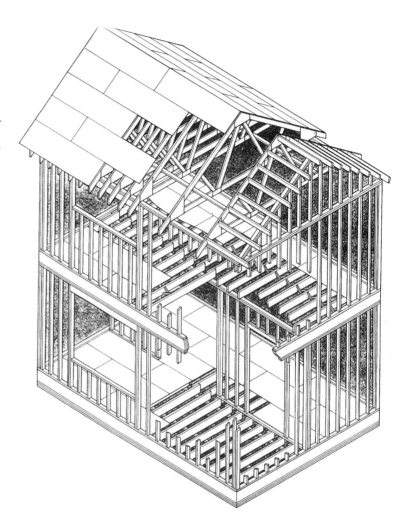

in mechanical sawmills. They were small and of uniform, rectangular cross-section and were held together by nails – another mass-produced product – rather than being connected by traditional carpentry joints. Lateral stability was provided by the boarding which formed the surfaces of the walls, floors and roof. The structural frameworks could be erected quickly and therefore cheaply by relatively unskilled labour. There were no complicated joints. The building type is still in use in the present day.

6.2.4 Timber skeleton-frame structures

In skeleton-frame structures the volume of structural material which is present is considerably smaller than in loadbearing-wall structures and the structural loads are concentrated into slender beams and columns. Stress levels are therefore high, and, because the strength of timber is only moderate compared to material such as steel, it is frequently considered unsuitable for skeleton frames. There is, however, a tradition of timber skeleton-frame architecture. This has received fresh impetus in the twentieth century by the development of laminated timber and other types of built-up beam which have allowed the creation of beams and columns with larger cross-sections than are possible with sawn timber. Skeleton frames in timber are characterised by elements with fairly large cross-section in relation to their length. The structure is normally allowed to dominate the architectural language (Fig. 6.15).

Fig. 6.15 Ragnitz Church Hall, Graz, Austria, Szyszkowitz and Kowalski, architects. The roof of this building is supported by a timber skeleton framework which has a significant architectural presence. [Photo: E. & F. McLachlan]

6.2.5 Twentieth-century developments

All three structure types which have been described in the preceding sections are in current use. Large timber trusses are used as the horizontal structures in wide-span buildings, timber wallframe houses account for a significant proportion of domestic buildings and the distinctive skeleton frame in timber is still part of the architectural vocabulary. The structures of the present day are significantly different from those of previous centuries due to recent developments in the technology of timber. In particular, modern timber structures are lighter and they are also more precisely crafted. The two very significant developments which have occurred in the twentieth century are the practice of stress grading and the evolution of greatly improved jointing techniques.

The stress grading system, which is described briefly in Section 6.5, overcomes one of the fundamental problems of timber which is the variability of the material. It allows the strength of timber elements to be specified within fairly precise limits. This in turn allows factors of safety which are used in the design calculations to be low. The material is therefore used efficiently.

As was seen in Section 6.2.3, traditional carpentry joints involve the cutting away of a considerable amount of material – sometimes as much as half the cross-section of a structural element. The joint was therefore the weakest part of the element and the resulting efficiency was very low. In modern timber joints the quantity of material which is removed is minimised. In addition, fastening techniques, based on improved glues and mechanical components, have been developed which minimise the concentration of stress which occurs at joints (see Section 6.6.6).

One of the main consequences of the availability of these improved jointing techniques has been the development of a wide variety of built-up elements such as laminated timber beams, portal frames and arches, ply-web beams and trussed rafters of various configurations. The availability of efficient types of joint has also allowed timber to compete effectively with steel for triangulated plane and space frameworks for roof structures.

6.3 The material, its properties and characteristics

6.3.1 Introduction
The technical advantages and disadvantages of timber in relation to other structural materials are reviewed in this section.

6.3.2 Advantages
Strength
Timber possesses tensile, compressive and flexural strength and is therefore suitable for all types of structural element.

Lightness
Timber is a lightweight material with a high ratio of strength to weight. It therefore produces lightweight structures with components which can be easily transported and handled on site.

Tractability
Timber can be easily cut and shaped with simple tools and the erection of timber structures is therefore straightforward. Other components can be easily attached to timber with simple fasteners, such as nails or screws, and this simplifies the detailing of timber buildings.

Performance in fire
Timber is a combustible material but the rate at which it is consumed in a fire is relatively low and it does not lose its structural properties when it is subjected to high temperatures. It therefore continues to function in a fire until the cross-section of elements become reduced to the point at which excessive stress occurs (other materials, particularly metals, lose their strength at relatively low temperatures). The performance of timber structures in fire is therefore good.

Durability
The constituents of timber are relatively stable chemically and the material does not suffer chemical degradation in environmental conditions (such as high humidity levels) which might prove detrimental to metals. It is, however, susceptible to insect infestation and fungal attack (see Section 6.4.4.4).

Appearance
Timber is a material which has a pleasing appearance which matures rather than deteriorates with age. It can therefore serve in the combined role of a structural material and a finishing material.

6.3.3 Disadvantages
Lack of strength
Although the strength-to-weight ratio of timber is high, its actual strength is low compared to other structural materials such as steel and

reinforced concrete. This restricts the size of span which can be achieved and the number of storeys which can be constructed in an all-timber structure.

Jointing difficulty
Although the material is tractable it is difficult to make joints in timber which have a good structural performance. Joints which are made with mechanical fasteners, such as nails, screws and bolts, suffer from the problems of stress concentration. They also tend to work loose due to the phenomenon of moisture movement (see Section 6.4.3). Although very significant advances have been made in recent decades in the development of new types of mechanical fastener, these problems have not been entirely overcome.

The development, in the twentieth century, of reliable glues has made possible the virtual elimination of both stress concentration and the development of looseness at joints. To achieve a satisfactory connection with glue, however, it is necessary that the contact surfaces should be very carefully prepared and also that the glue should be properly cured.[4] Both of these are difficult to achieve on site and this requires that gluing be carried out under factory conditions only. The size of component which can be assembled with glued joints is therefore restricted by considerations of transport.

Component size
The sizes of individual timber planks or boards are obviously determined by the size of available trees. A consequence of the depletion of the world's resources of timber is that lengths greater than around 6 m and cross-sectional dimensions in excess of 300 mm by 100 mm are difficult to obtain in most of the species which are used for structural purposes. This, together with the jointing problems described above and the low relative strength, restricts the overall size of the structures which can be constructed in timber in the present day.

Susceptibility to rot and decay
Timber components are susceptible to various kinds of infestation, notably by fungi and insects. The likelihood of fungal attack is minimised if the timber is kept dry and this can be achieved through suitable detailing of the structure. Insect attack is inhibited by application of insecticides but the effectiveness of these is limited.

Variability
Timber, being a natural material, exhibits considerable variability in its properties. Different species produce timbers with quite different properties and even within a single species the variability in properties such as strength, elasticity, durability and appearance can be high. The problem has been overcome by the adoption of a grading system for commercial timber in which individual planks and boards are inspected and placed into categories according to their particular characteristics. Timber which is used in construction is specified by grade and this ensures that its properties can be relied upon to be within known limits.

6.3.4 Conclusion
To sum up, timber is a material which offers the designers of buildings a combination of properties which allow the creation of light-weight structures which are simple to construct. Its relatively low strength, the small sizes of the basic components and the difficulties associated with achieving good structural joints tend to limit the size of structure which is possible, however, and the majority of timber structures are small in scale with short spans and a small number of storeys. Currently, the most common application of timber in architecture is in domestic building where it is used as a primary structural material, either to form the entire structure for a building, as in timber wallframe construction (also called timber-frame construction) (Fig. 6.14), or to make the horizontal elements in loadbearing-masonry structures (Fig. 6.2).

The high ratio of strength to weight which is found in timber also makes it particularly

---

4 See Gordon, *The New Science of Strong Materials*, Harmondsworth, 1968.

suitable for structures of moderately large span, especially if the level of imposed load is low. Timber is therefore used for the construction of large, single-storey enclosures and for medium- to long-span roof structures.

6.4 Properties of timber

6.4.1 Introduction
In this short section only the properties of timber which are relevant to its use as a structural material are reviewed. These are the mechanical properties (i.e. strength and elasticity), fire resistance and durability. Two introductory sections, in which the internal structure of timber and the phenomenon of moisture movement are described briefly, are also included. More detailed accounts of all of these topics will be found in Desch, H. E., *Timber*, 6th ed. (revised Dinwoodie), London, 1981.

6.4.2 Internal structure
The parts of the tree which are used as the source of structural timber are the trunk and sometimes the major branches. These, in common with other parts of a tree, are composed of cells; in the trunk these are long and thin and are aligned parallel to its direction. The mechanical strength of timber is derived mainly from the fibrous tissue which forms the walls of the cells together with deposits within the cells.

The most active part of a living tree trunk is the layer which lies immediately under the bark. This is called the cambium layer and it is here that growth by cell division occurs. In seasonal climates the rate of growth varies throughout the year. It takes place mostly in the spring and summer and is more rapid in spring than in summer. Spring wood is softer than summer wood, and this difference accounts for the annual rings which are a prominent feature of the cross-section of a tree trunk and which are responsible for the phenomenon of 'grain'. The denser, harder parts of the grain are formed by summer wood and the softer parts by spring wood.

Trees may be classified into two types, narrow-leaved trees and broad-leaved trees. Narrow-leaved trees are coniferous and mostly evergreen – an exception is larch; broad-leaved trees are mainly deciduous – an exception to this is holly. Many differences exist between the physiologies and anatomical structures of these two types of tree. The most significant of them is that the narrow-leaved species tend to grow much faster and to produce timbers which are less dense and less strong than the broad-leaved species.

Commercial timbers are subdivided into the two broad categories of softwoods and hardwoods, and these correspond approximately to the botanical classifications. The softwoods are derived from the narrow-leaved species and the hardwoods from the broad-leaved species. The terms are misleading, however, because in some cases they do not correspond to the physical properties of the timber. Balsa wood, for example, which is one of the softest of timbers, is classified as a hardwood because it is derived from a broad-leaved tree. In general, however, the coniferous species from which the softwoods are derived are fast-growing and produce timbers with a high proportion of spring wood. They are therefore less dense and less rigid than the hardwood timbers which are derived from slow-growing, broad-leaved varieties.

6.4.3 Moisture content and moisture movement
The moisture content of a specimen of timber is defined as the ratio of the weight of water which it contains to its dry weight. It is always expressed as a percentage. In the living tree the moisture content is around 150%. After the tree is cut down, most of the water evaporates and the timber eventually reaches an 'equilibrium moisture content' whose exact value depends on the species and on the conditions of the environment in which it is placed; it depends principally on temperature and relative humidity and is usually in the region of 15 to 20%.

When timber dries out following the cutting down of a tree, moisture is lost initially from the interiors of the cells and the only physical change which this causes is a reduction in density. At a moisture content of around 30% (the exact figure depends on the species), all of the moisture in the cell interiors will have been lost and only the water in the cell walls will remain. This condition is known as the 'fibre saturation point' of the timber. Further reduction in the moisture content causes water to be lost from the cell walls and this gives rise to shrinkage as well as to loss of weight. The shrinkage is greater in the across-grain directions than in the along-grain direction and, of the across-grain directions, it is greater tangentially than radially (Fig. 6.16). The shrinkage is reversible; if the equilibrium moisture content of a timber becomes higher, due to a change in environmental conditions, then the timber will absorb water from the atmosphere and a corresponding expansion will occur. The phenomenon is known as 'moisture movement' and the extent to which it occurs depends on the species.

The greatest amount of 'movement' takes place during the initial drying process which must be regulated carefully because differential shrinkage, due to variations in the rates at which moisture is lost from different parts, could cause damage in the form of cracks and warping. The controlled drying out of timber after felling is called 'seasoning'. It involves the adoption of measures to restrict the rate at which moisture is lost from the timber and to restrain the latter physically so as to prevent excessive deformation from occurring. Various techniques for seasoning have been developed but most timber which is used in building is either 'air-seasoned' or 'kiln-seasoned'. In air-seasoning the timber is stacked in well-ventilated sheds and allowed to dry out under atmospheric conditions; in kiln-seasoning the timber is dried more quickly under conditions of close environmental control. In both cases the stacking arrangement which is adopted is designed both to allow air to circulate freely between timbers, so as to promote drying, and to restrain the timber to prevent it from warping. A certain amount of deterioration in the quality of the timber, due to the formation

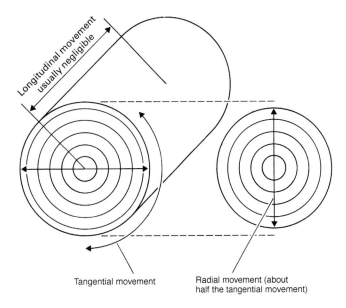

Fig. 6.16 The shrinkage which occurs to timber during the initial drying out is greater in the tangential than in the radial direction.

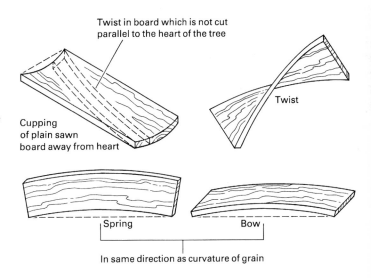

Fig. 6.17 Examples of seasoning defects.

of cracks and to warping (Fig. 6.17), is inevitable during seasoning, however, because differences in the rates at which moisture is lost from different parts of each timber cannot

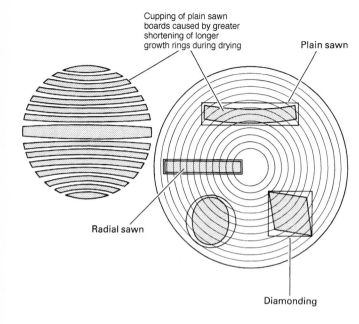

Cupping of plain sawn
boards caused by greater
shortening of longer
growth rings during drying

Plain sawn

Radial sawn

Diamonding

Fig. 6.18 Seasoning defects which occur due to the difference between the rate of radial and tangential shrinkage.

be entirely eliminated. The extent and incidence of seasoning defects is reduced if small cross-sections (i.e. cross-sections in which one of the dimensions is 75 mm or less) are used; these factors are also affected by the original location and orientation of a plank within the tree (Fig. 6.18)

The equilibrium moisture content of the timber in a building depends on the internal environment of the building. In the UK it is usually around 15% and is sometimes less in buildings in which high temperatures are maintained. Air-seasoned timber is supplied with a moisture content of around 22% and will therefore suffer further drying and shrinkage after installation in a building. Kiln-seasoned timber can be supplied with a lower moisture content and is less likely to suffer a large shift in moisture content after installation. All timber is likely to undergo small shifts in moisture content, with corresponding dimensional changes, during the life of a building, because of changes in the environmental conditions.

6.4.4 Mechanical properties

6.4.4.1 Introduction
Due to the fibrous nature of the material and the phenomenon of grain, the strength of timber and its behaviour in respect of deformation are non-isotropic (not the same in all directions).[5] Many of the features of the mechanical properties of timber can in fact be understood by imagining its internal structure to be similar to that of a bundle of drinking straws.

6.4.4.2 Strength
In the direction parallel to the grain the strength of timber is slightly less in compression than it is in tension. This is due to the tendency for the cell walls, which are aligned parallel to the grain, to buckle under the action of compressive loads. The strength at right angles to the grain is considerably less than that parallel to the grain. This is due to separation of the fibres when a tensile load is applied and crushing of the cells when a compressive load is applied. A number of different values are therefore normally quoted for the strength of a timber depending on the direction in which a load is appiied with respect to the grain.

The grade stress values which are given in the British Standard (BS 5268), are reproduced here in Table 6.1. These are derived from the results of tests carried out on structural-sized elements rather than small specimens of the species concerned, but they are subject to modification to allow for differences between the test conditions and the conditions which are likely to prevail in a real structure in service. Examples of the latter are the duration of the load (the ability of timber to sustain a load depends on the length of time for which it is applied), the moisture content of the timber, the particular dimensions of the cross-section of the element and a number of other factors.

---

5 The strength and elasticity of a specimen of timber depend on the orientation of the load with respect to the grain. Strength and modulus of elasticity are greater parallel to the grain than normal to the grain.

Table 6.1 Grade stresses for softwood timbers (after BS 5268)

Standard	Grade	Bending parallel to grain	Tension parallel to grain	Compression Parallel to grain	Perpendicular to grain	Shear parallel to grain	Modulus of elasticity Mean	Minimum
		N/mm²	N/mm²	N/mm²	N/mm²	N/mm²	N/mm²	N/mm²
Redwood/whitewood	SS/MSS	7.5	4.5	7.9	2.1	0.82	10500	7000
(imported) and	GS/MGS	5.3	3.2	6.8	1.8	0.82	9000	6000
Scots pine,	M75	10.0	6.0	8.7	2.4	1.32	11000	7000
(British grown)	M50	6.6	4.0	7.3	2.1	0.82	9000	6000
Corsican pine	SS/MSS	7.5	4.5	7.9	2.1	0.82	9500	6500
(British grown)	GS/MGS	5.3	3.2	6.8	1.8	0.82	8000	5000
	M75	10.0	6.0	8.7	2.4	1.33	10500	7000
	M50	6.6	4.0	7.3	2.0	0.83	9000	5500
Sitka spruce and	SS/MSS	5.7	3.4	6.1	1.6	0.64	8000	5000
European spruce	GS/MGS	4.1	2.5	5.2	1.4	0.64	6500	4500
(British grown)	M75	6.6	4.0	6.4	1.8	1.02	9000	6000
	M50	4.5	2.7	5.5	1.6	0.64	7500	5000
Douglas fir	SS/MSS	6.2	3.7	6.6	2.4	0.88	11000	7000
(British grown)	GS/MGS	4.4	2.6	5.6	2.1	0.88	9500	6000
	M75	10.0	6.0	8.7	2.9	1.41	11000	7500
	M50	6.6	4.0	7.3	2.4	0.88	9500	6000
Larch	SS	7.5	4.5	7.9	2.1	0.82	10500	7000
(British grown)	GS	5.3	3.2	6.8	1.8	0.82	9000	6000
Parana pine	SS	9.0	5.4	9.5	2.4	1.03	11000	7500
(imported)	GS	6.4	3.8	8.1	2.2	1.03	9500	6000
Pitch pine	SS	10.5	6.3	11.0	3.2	1.16	13500	9000
(Caribbean)	GS	7.4	4.4	9.4	2.8	1.16	11000	7500
Western red cedar	SS	5.7	3.4	6.1	1.7	0.63	8500	5500
(imported)	GS	4.1	2.5	5.2	1.6	0.63	7000	4500
Douglas fir-larch	SS	7.5	4.5	7.9	2.4	0.85	11000	7500
(Canada)	GS	5.3	3.2	6.8	2.2	0.85	10000	6500
Douglas fir-larch	SS	7.5	4.5	7.9	2.4	0.85	11000	7500
(USA)	GS	5.3	3.2	6.8	2.2	0.85	9500	6000
Hem-fir	SS/MSS	7.5	4.5	7.9	1.9	0.68	11000	7500
(Canada)	GS/MGS	5.3	3.2	6.8	1.7	0.68	9000	6000
	M75	10.0	6.0	9.3	2.4	1.13	12000	8000
	M50	6.6	4.0	7.7	2.1	0.71	10500	7000
Hem-fir	SS	7.5	4.5	7.9	1.9	0.68	11000	7500
(USA)	GS	5.3	3.2	6.8	1.7	0.68	9000	6000
Spruce-pine-fir	SS/MSS	7.5	4.5	7.9	1.8	0.68	10000	6500
(Canada)	GS/MGS	5.3	3.2	6.8	1.6	0.68	8500	5500
	M75	9.7	5.8	8.5	2.1	1.10	10500	7000
	M50	6.2	3.7	7.1	1.8	0.68	9000	5500
Western whitewoods	SS	6.6	4.0	7.0	1.7	0.66	9000	6000
(USA)	GS	4.7	2.8	6.0	1.5	0.66	7500	5000
Southern pine	SS	9.6	5.8	10.2	2.5	0.98	12500	8500
(USA)	GS	6.8	4.1	8.7	2.2	0.98	10500	7000

The allowable stress which is used in the design of a timber element is the grade stress multiplied by one or more stress modification factors, which allow for the effects noted above.

Because timber is much stronger parallel to the grain than in other directions most timber components are designed so that the stress which results from the principal load occurs parallel to the grain.

6.4.4.3 Fire resistance

Although timber is combustible its performance in fire is good, compared to that of other materials such as steel, aluminium or plastic. There are two reasons for this. Firstly, the structural performance of timber is more-or-less independent of temperature (within the range of temperatures likely to be experienced in a building fire) so that a timber structure will continue to support load during a fire until the cross-sections of the elements are reduced by combustion to the point at which the stresses in them become excessive. It does not collapse as a result of the effects of high temperature alone. Secondly, the rate at which timber is consumed in a fire is low and relatively constant. It is therefore possible to design a timber structure to withstand the effects of fire for a predetermined period, by oversizing the elements to allow for the material which will be consumed in that period. This method of design for fire resistance is known as the 'sacrificial timber' or 'residual section' design method. Charring rates for timber are given in Table 6.2.

6.4.4.4 Durability

Timber structures are likely to suffer from two types of decay: insect attack and fungal attack. Fungal attack is the more destructive of these, but this will only occur if the moisture content of the timber is greater than 20% and can be prevented by suitable design. Timber structures are therefore detailed so that direct contact with the ground is avoided and in such a way as to minimise the risk of condensation forming on timber elements. Features which will allow water to collect and lie on timber are also avoided. The likelihood of fungal and insect attack can also be reduced by impregnating the timber with various preservative materials.

6.5 Grading of timber

The properties of timber can exhibit considerable variability, even within samples from a single species, and where the material is used for structural purposes it is necessary that this variability should be quantified. The system which is adopted for dealing with variability is one of grading. Processed timber is placed into grades, which are simply categories, and individual pieces of timber must comply with certain performance requirements to qualify for inclusion in a particular grade. All timber is therefore inspected or tested at some stage during its processing and allocated a grade depending on its performance. The user of timber specifies the grade which must be used for a particular component as well as its species, and can then be confident that its properties will be within defined limits.

Most timber is graded in its country of origin and, while the basic principles which are adopted for grading are similar for all major producing countries, there are differences between countries in the details of the methods. For softwoods, which account for most structural timber in contemporary building practice, most countries operate what is

Table 6.2 Charring rates of timber

	Loss of timber from one face of element (mm)				
	Minutes				
	15	30	45	60	90
Softwood including laminated timber but excluding western red cedar	10	20	30	40	60
Western red cedar	12.5	25	37.5	50	62.5
Hardwoods	7.5	15	22.5	30	45

Table 6.3 Relationship of 'appearance' grades for softwood timber

Norway, Sweden, Finland, Poland and Eastern Canada		I, II, III, IV unsorted	V	VI
USSR		I, II, III unsorted	IV	V
Brazil	No. I and No. 2			
British Columbia and Pacific Coast of North America (R list)	No. I clear No. 2 clear No. 3 clear	Select merchantable No. 1 merchantable	No. 2 merchantable	No. 3 common
UK BS 3819	I Clear	I, II	III	IV

known as the 'appearance' or 'defect' system of grading in which limits are set for the sizes and frequency of occurrence of various types of defect such as knots, grain (its width, straightness, angle of inclination to the principal direction of the member, etc.) and the defects which arise from conversion[6] and seasoning, such as wane, splits, twist, cupping and bowing (Fig. 6.17). Each piece of timber is simply inspected and placed in a grade according to the sizes and frequency of defects which it contains. For each grade the limits on the sizes of defects are related to the dimensions of the piece of timber being considered. Thus, large defects in a large plank can result in it being placed in the same grade as a smaller plank with smaller defects. If a large plank is sawn up into two or more smaller planks the timber must be regraded. Most countries have adopted either four, five or six grades. Table 6.3 shows the relationship between the grades which are used by a number of major sources of structural timber.

The appearance or defects method of grading is not entirely satisfactory as a means of classifying structural timber because the appearance of the surface of a specimen is not a reliable enough guide to its strength properties. For this reason a system of stress-grading has been adopted in the UK, and all timber which is used in structures which have been designed to comply with BS 5268 must be graded by this method, even though it may already have been graded by the appearance method. The British Standard in which the stress grading procedures are specified is BS 4978.

The stress-grading system does in fact make use of visual criteria but these are based on the effects which knots and other defects have on the cross-section of the member rather than on the appearance of its surface and therefore give a more realistic assessment of the effect of defects on strength. Two visual stress grades, general structural (GS) and special structural (SS) are specified for each species of sawn timber and three visual stress grades are specified for laminated timber (LA, LB and LC). The visual stress-grading system is now being used in conjunction with 'machine-grading' in which the timber is passed through a device which imposes a bending load on it and simultaneously measures the resulting deflection.

---

6 Conversion is the name given to the process by which felled timber in the form of tree trunks is sawn up into planks.

Table 6.4 Basic sawn sizes of softwood (from BS 4471 Part 1: 1978)

Thickness (mm)	Width (mm)								
	75	100	125	150	175	200	225	250	300
16	X	X	X	X					
19	X	X	X	X					
22	X	X	X	X					
25	X	X	X	X	X	X	X	X	X
32	X	X	X	X	X	X	X	X	X
36	X	X	X	X					
38	X	X	X	X	X	X	X		
44	X	X	X	X	X	X	X	X	X
47	X	X	X	X	X	X	X	X	X
50	X	X	X	X	X	X	X	X	X
63		X	X	X	X	X	X		
75		X	X	X	X	X	X	X	X
100		X			X		X		X
150				X		X			X
200						X			
250								X	
300									X

This system is based on the fact that the bending deformation which occurs to a specimen of timber is a reliable guide to its strength.

The stress values which are quoted here in Table 6.1 are the grade stresses which are currently used in the UK. They are reproduced from BS 5268. The prefix M indicates that the material must be machine graded.

6.6 Timber components

6.6.1 Solid timber

6.6.1.1 Sawn timber and its derivatives
The simplest timber components are solid elements of rectangular cross-section on which no finishing was carried out after conversion. These are referred to as 'sawn-timber' elements. The range of cross-sectional sizes (basic sawn sizes) and lengths in which softwood timber is available in the UK are given in Tables 6.4 and 6.5 and the dimensional properties of their cross-sections are given in Table 6.6. Note that the terminology which is used to describe the cross-section is to call the smaller of the two dimensions the thickness and the larger the width.

Basic sawn-timber elements tend to have a slightly irregular geometry and in situations where this is undesirable, such as in trusses in which the elements are joined by gusset plates and must therefore be of equal thickness, the timber is 'regularised'. This is a machine process by which the thickness and/or width of an element is made uniform throughout its length. It is normally carried out on the critical dimension only: that is on the thickness, for elements which are to be used in trusses, and on the width, for elements, such as floor joists or wall studs, where this is the critical dimension. Regularising involves a reduction of around 3 mm in the dimension which is affected.

Table 6.5 Basic lengths of sawn softwood (m) (From BS 4471 Part 1: 1978)

1.80	2.10	3.00	4.20	5.10	6.00	7.20
	2.40	3.30	4.50	5.40	6.30	
	2.70	3.60	4.80	5.70	6.60	
		3.90			6.90	

Table 6.6 Geometrical properties of sawn softwoods

Basic size*	Area	Section modulus		Second moment of area		Radius of gyration	
		About x–x	About y–y	About x–x	About y–y	About x–x	About y–y
mm	$10^3 mm^2$	$10^3 mm^3$	$10^3 mm^3$	$10^6 mm^4$	$10^6 mm^4$	mm	mm
36 × 75	2.70	33.8	16.2	1.27	0.292	21.7	10.4
36 × 100	3.60	60.0	21.6	3.00	0.389	28.9	10.4
36 × 125	4.50	93.8	27.0	5.86	0.486	36.1	10.4
36 × 150	5.40	135	32.4	10.1	0.583	43.3	10.4
38 × 75	2.85	35.6	18.1	1.34	0.343	21.7	11.0
38 × 100	3.80	63.3	24.1	3.17	0.457	28.9	11.0
38 × 125	4.75	99.0	30.1	6.18	0.572	36.1	11.0
38 × 150	5.70	143	36.1	10.7	0.686	43.3	11.0
38 × 175	6.65	194	42.1	17.0	0.800	50.5	11.0
38 × 200	7.60	253	48.1	25.3	0.915	57.7	11.0
38 × 225	8.55	321	54.2	36.1	1.03	65.0	11.0
44 × 75	3.30	41.3	24.2	1.55	0.532	21.7	12.7
44 × 100	4.40	73.3	32.3	3.67	0.710	28.9	12.7
44 × 125	5.50	115	40.3	7.16	0.887	36.1	12.7
44 × 150	6.60	165	48.4	12.4	1.06	43.3	12.7
44 × 175	7.70	225	56.5	19.7	1.24	50.5	12.7
44 × 200	8.80	293	64.5	29.3	1.42	57.7	12.7
44 × 225	9.90	371	72.6	41.8	1.60	65.0	12.7
44 × 250	11.0	458	80.7	57.3	1.77	72.2	12.7
44 × 300	13.2	660	96.8	99.0	2.13	86.6	12.7
47 × 75	3.53	44.1	27.6	1.65	0.649	21.7	13.6
47 × 100	4.70	78.3	36.8	3.92	0.865	28.9	13.6
47 × 125	5.88	122	46.0	7.65	1.08	36.1	13.6
47 × 150	7.05	176	55.2	13.2	1.30	43.3	13.6
47 × 175	8.23	240	64.4	21.0	1.51	50.5	13.6
47 × 200	9.40	313	73.6	31.3	1.73	57.7	13.6
47 × 225	10.6	397	82.8	44.6	1.95	65.0	13.6
47 × 250	11.8	490	92.0	61.2	2.16	72.2	13.6
47 × 300	14.1	705	110	106	2.60	86.6	13.6
50 × 75	3.75	46.9	31.3	1.76	0.781	21.7	14.4
50 × 100	5.00	83.3	41.7	4.17	1.04	28.9	14.4
50 × 125	6.25	130	52.1	8.14	1.30	36.1	14.4
50 × 150	7.50	188	62.5	14.1	1.56	43.3	14.4
50 × 175	8.75	255	72.9	22.3	1.82	50.5	14.4
50 × 200	10.0	333	83.3	33.3	2.08	57.7	14.4
50 × 225	11.3	422	93.8	47.5	2.34	65.0	14.4
50 × 250	12.5	521	104	65.1	2.60	72.2	14.4
50 × 300	15.0	750	125	113	3.13	86.6	14.4
63 × 100	6.30	105	66.2	5.25	2.08	28.9	18.2
63 × 125	7.88	164	82.7	10.3	2.60	36.1	18.2
63 × 150	9.45	236	99.2	17.7	3.13	43.3	18.2
63 × 175	11.0	322	116	28.1	3.65	50.5	18.2
63 × 200	12.6	420	132	42.0	4.17	57.7	18.2
63 × 225	14.2	532	149	59.8	4.69	65.0	18.2
75 × 100	7.50	125	93.8	6.25	3.52	28.9	21.7
75 × 125	9.38	195	117	12.2	4.39	36.1	21.7
75 × 150	11.3	281	141	21.1	5.27	43.3	21.7
75 × 175	13.1	383	164	33.5	6.15	50.5	21.7
75 × 200	15.0	500	188	50.0	7.03	57.7	21.7
75 × 225	16.9	633	211	71.2	7.91	65.0	21.7
75 × 250	18.8	781	234	97.7	8.79	72.2	21.7
75 × 300	22.5	1130	281	169	10.5	86.6	21.7
100 × 100	10.0	167	167	8.33	8.33	28.9	28.9
100 × 150	15.0	375	250	28.1	12.5	43.3	28.9
100 × 200	20.0	667	333	66.7	16.7	57.7	28.9
100 × 250	25.0	1040	417	130	20.8	72.2	28.9
100 × 300	30.0	1500	500	225	25.0	86.6	28.9
150 × 150	22.5	563	563	42.2	42.2	43.3	43.3
150 × 200	30.0	1000	750	100	56.3	57.7	43.3
150 × 300	45.0	2250	1130	338	84.4	86.6	43.3
200 × 200	40.0	1330	1330	133	57.7	57.7	57.7
250 × 250	62.5	2600	2600	326	326	72.2	72.2
300 × 300	90.0	4500	4500	675	675	86.6	86.6

*Basic size measured at 20% moisture content.

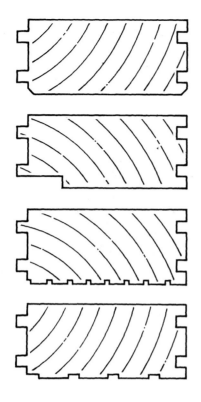

Where a high quality of finish is required, timber elements are planed on all four sides (planed-all-round). This results in a larger reduction in the cross-sectional dimensions than occurs with regularising. Where a timber element has been regularised or planed-all-round the structural calculations are based on the modified dimensions.

6.6.1.2 Solid timber decking

Timber decking is machine-finished timber of rectangular cross-section with edges which are tongued-and-grooved so that boards can be assembled into a flat deck in which adjacent elements give each other mutual support (Fig. 6.19). Traditionally, this was produced in small thicknesses, to span short distances of around 300 mm to 450 mm, and was used in conjunction with sawn-timber beams (joists) to form floor or roof decks. The range of thicknesses has been extended in recent years to allow wider spacing of beams. (See Table 6.7 and Fig. 6.20.)

Fig. 6.19 The tongues and grooves in manufactured decking elements allow concentrated loads to be distributed.

Table 6.7 Span–load table for tongued and grooved solid timber decking

(Based upon S.S. grade E Whitewood)

Load kN m²	Single span (m)				Continuous span (m)			
	38 mm (33)	50 mm (45)	63 mm (58)	75 mm (70)	38 mm (33)	50 mm (45)	63 mm (58)	75 mm (70)
1.0	2.15	2.94	3.79	4.57	2.89	3.94	5.08	6.12
1.1	2.08	2.85	3.67	4.43	2.79	3.82	4.92	5.93
1.2	2.09	2.77	3.57	4.30	2.72	3.71	4.78	5.76
1.3	1.97	2.69	3.47	4.19	2.64	3.61	4.65	5.61
1.4	1.92	2.63	3.39	4.09	2.58	3.52	4.54	5.47
1.5	1.88	2.57	3.31	3.99	2.52	3.44	4.43	5.35
1.6	1.84	2.51	3.24	3.91	2.46	3.37	4.34	5.23
1.7	1.80	2.46	3.17	3.83	2.41	3.30	4.25	5.13
1.8	1.77	2.42	3.11	3.76	2.37	3.24	4.17	5.03
1.9	1.74	2.37	3.06	3.69	2.33	3.18	4.09	4.94
2.0	1.71	2.33	3.01	3.63	2.29	3.12	4.02	4.86
2.5	1.58	2.16	2.79	3.37	2.12	2.90	3.73	4.51
3.0	1.49	2.04	2.62	3.17	1.99	2.73	3.51	4.24
3.5	1.41	1.93	2.49	3.01	1.89	2.59	3.34	4.03
4.0	1.35	1.85	2.38	2.88	1.81	2.47	3.19	3.85
4.5	1.30	1.78	2.29	2.76	1.74	2.38	3.06	3.70
5.0	1.25	1.72	2.21	2.67	1.68	2.30	2.96	3.57
5.5	1.21	1.66	2.14	2.58	1.63	2.22	2.86	3.46
6.0	1.18	1.61	2.08	2.51	1.58	2.16	2.78	3.36

Fig. 6.20
Tongued-and-grooved decking elements are used here in conjunction with laminated timber frames. The use of thick cross-sections for the decking elements allows the elimination of secondary structural elements such as purlins.

6.6.2 Composite boards

6.6.2.1 Introduction
Composite boards are manufactured products composed of wood and glue (Fig. 6.21). They are intended to exploit the advantages of timber while at the same time minimising the effects of its principal disadvantages, which are variability, dimensional instability, restrictions in the sizes of individual components and anisotropic behaviour.

There are two basic types of composite board: laminated boards (also called plywoods), in which the constituent parts are relatively large and are either veneers or strips of wood, and particle boards, in which the constituent pieces of timber are very small fibres or particles.

6.6.2.2 Laminated boards
The components of laminated boards are assembled by gluing such that the direction of the grain is different in adjacent layers. This, together with the high level of glue impregnation, reduces variability and improves dimensional stability in the plane of the board. It

also reduces the tendency of the board to split in the vicinity of nails and screws.

Laminated boards are subdivided into two categories: veneer plywood and core plywood. Veneer plywood is defined as plywood in which all the plies are made up of veneers, up to 7 mm thick, orientated with their grain parallel to the surfaces of the panel. Core plywood is plywood which has a relatively thick core and which may contain layers with grain normal to the plane of the board. Batten board, block-board and laminboard are included in this second group (Fig. 6.21).

The plywood which has been most extensively used as a structural material is veneer plywood. A large number of types of this are produced by the timber industry but few are considered suitable for structural use. The principal structural plywoods which are used in the UK are listed in BS 5268. These are usually available in sheets measuring 2440 mm by 1220 mm and in a range of thicknesses and surface qualities. The good in-plane rigidity and out-of-plane bending strength of veneer plywood make it suitable for a variety of structural applications, usually in some form of

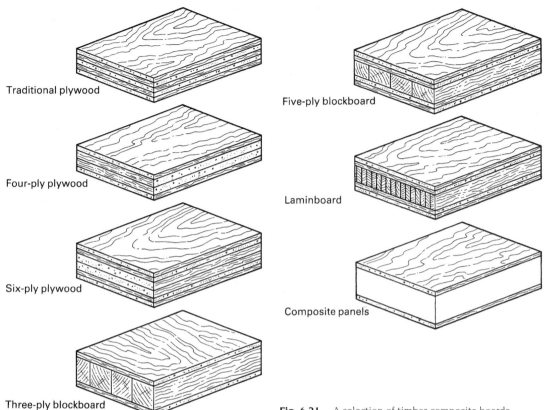

Traditional plywood

Four-ply plywood

Six-ply plywood

Three-ply blockboard

Five-ply blockboard

Laminboard

Composite panels

Fig. 6.21 A selection of timber composite boards.

composite construction. It is ideal for shear panels such as the webs of built-up-beams (Fig. 6.25) or the skins of wall or floor panels which are used for diaphragm bracing. It is also used as the flanges in stressed-skin panel structure (Fig. 6.40).

6.6.2.3 Particle boards
Particle boards have traditionally been little used as structural components and no data for them are given in BS 5268. There is an increasing tendency for these to be used in the structural role, however, particularly as a substitute for tongued-and-grooved floor and roof decking.

6.6.3 Built-up-beams

6.6.3.1 Introduction
Built-up-beams are components in which a number of separate planks or boards are

assembled, by gluing, nailing, screwing, bolting or a combination of these, to form beams of composite cross-section. Various types of built-up-beam are produced and these are all intended to exploit the good structural properties of timber and to overcome the size restriction which is inherent in the use of solid sawn-timber elements. They are normally used in forms of construction which are similar to those which are suitable for sawn timber (i.e. as joists in the traditional forms of floor and roof structures – see Fig. 6.41) and they allow longer spans to be achieved than are possible with solid sawn-timber. Some of the more commonly used forms of built-up-beams are reviewed briefly here.

6.6.3.2 Laminated timber
Laminated timber, which is also called 'gluelam', is a product in which elements with large rectangular cross-sections are built up by

Fig. 6.22 Typical cross-section of a laminated timber element [Photo: TRADA].

gluing together smaller solid timber members of rectangular cross-section, usually either 38 mm or 50 mm in thickness (Fig. 6.22). The direction of the grain is the same for all laminations and is usually parallel to the axis of the element. The obvious advantage of the process is that it allows the manufacture of solid elements with much larger cross-sections than are possible in plain sawn timber. Very long elements are also possible because the constituent boards are joined end-to-end by

means of finger joints (Fig. 6.23). The laminating process also allows the construction of elements which have a curved profile (Figs 6.3, 6.20 and 6.24).

The general quality and strength of laminated timber is higher than that of sawn-timber for two principal reasons. Firstly, the use of basic components which have small cross-sections allows more effective seasoning, with fewer seasoning defects, than can be achieved with large sawn-timber elements. Secondly, the use of the finger joint, which causes a minimal reduction in strength in the constituent boards, allows any major defects which are present in these to be cut out. The allowable stresses of laminated timber are therefore higher than for equivalent sawn timber. Laminated timber elements are produced in three standards of finish: architectural (machined, sanded and free from blemishes), industrial (machined and sanded but with blemishes permitted) and economy (untreated after manufacture), but there is no structural difference between these categories.

6.6.3.3 Plyweb beams
Plyweb beams are built-up-beams of rectangular-box or I-cross-section in which solid timber flanges are used in conjunction with a web of

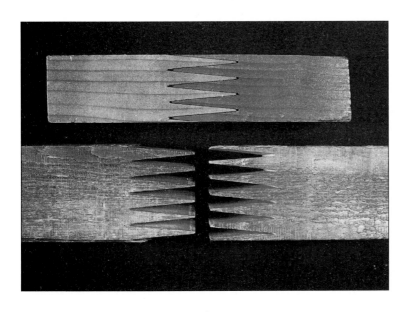

Fig. 6.23 The glued finger joint is an essential part of the technology of laminated timber. It allows the creation of very long sub-elements from which major defects, such as knots, have been removed [Photo: TRADA].

Fig. 6.24 St Joseph's Chapel, Ipswich. J. Powlesland, structural engineer. The timber roof structure here is based on large cross-section laminated timber elements, which are used in conjunction with laminated timber secondary elements and solid timber decking [Photo: Sidney Baynton]

veneer plywood (Fig. 6.25). Plyweb beams are normally considerably cheaper than gluelam equivalents. They are used in similar types of construction to solid sawn-timber beams but they allow wider spacings (around 1.3 m) and longer spans (up to 20 m) to be achieved.

6.6.3.4 Other forms of built-up-beams
A number of other types of built-up-beam are produced in timber. One example is the lattice-type beam, which is illustrated in Fig. 6.26.

6.6.4 Trussed rafters

Trussed rafters are lightweight timber trusses, Fig. 6.27; they are usually factory-made assemblies of small, sawn-timber sections, regularised on thickness and jointed with punched-metal-plate fasteners, but they can be assembled on site with bolted joints. Typical individual element sizes are 38 mm × 75 mm for short-span versions (6 m) and 50 mm × 100 mm for the larger versions (10 m). Most manufacturers produce a range of sizes in 100 mm increments of span and 2.5° increments of pitch between 15° and 35°.

Trussed rafters are designed to be used in pitched roof construction and spaced at close centres (around 600 mm) so as to carry the roof cladding without the need for a secondary structural system, other than tile battens or sarking boards. They therefore perform a similar role to that of the rafters in traditional

Fig. 6.25 Plyweb beams are built-up-beams with sawn-timber flanges and plywood webs. Composite elements such as this are used to extend the span range of traditional joisted forms of construction. Note the web stiffeners in the beams here which are positioned at the support points where shear loads are high.

Fig. 6.26 Built-up lattice beams. Proprietary components such as these are used, like plyweb beams, to extend the span ranges of traditional forms of timber construction.

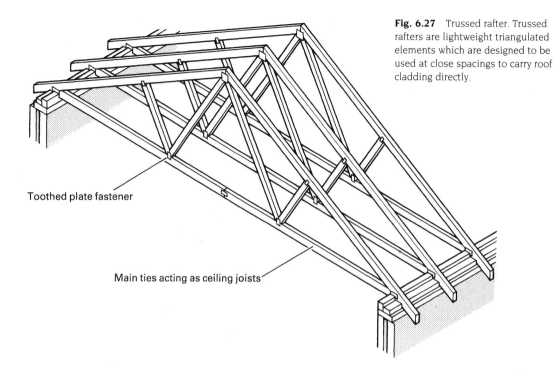

Fig. 6.27 Trussed rafter. Trussed rafters are lightweight triangulated elements which are designed to be used at close spacings to carry roof cladding directly.

Toothed plate fastener

Main ties acting as ceiling joists

forms of roof construction, from which they derive their name, but they allow a much more economical use of timber to be made due to the use of the triangulated form. They are generally considered to be most suitable for spans in the range 5 m to 11 m. Typical spans for which trussed rafters are suitable are listed in Table 6.8.

6.6.5 Large one-off trusses

Large trusses are used as the horizontal elements in post-and-beam structures of both the skeleton-frame and the loadbearing-wall types. They are normally only specified for situations where very high strength or rigidity is required, either because a long span must be achieved or because heavy loads are carried

Table 6.8 Typical spans for fully triangulated trussed rafters

Rafter size					Span Pitch (degrees)				
	15	17½	20	22½	25	27½	30	32½	35
mm	m	m	m	m	m	m	m	m	m
38 × 75	6.03	6.16	6.29	6.41	6.51	6.60	6.70	6.80	6.90
38 × 100	7.48	7.67	7.83	7.97	8.10	8.22	8.34	8.47	8.61
38 × 125	8.80	9.00	9.20	9.37	9.54	9.68	9.82	9.98	10.16
44 × 75	6.45	6.59	6.71	6.83	6.93	7.03	7.14	7.24	7.35
44 × 100	8.05	8.23	8.40	8.55	8.68	8.81	8.93	9.09	9.22
44 × 125	9.38	9.60	9.81	9.99	10.15	10.31	10.45	10.64	10.81
50 × 75	6.87	7.01	7.13	7.25	7.35	7.45	7.53	7.67	7.78
50 × 100	8.62	8.80	8.97	9.12	9.25	9.38	9.50	9.66	9.80
50 × 125	10.01	10.24	10.44	10.62	10.77	10.94	11.00	11.00	11.00

due to the infrequent spacing of the primary structural elements. So far as the overall geometry of trusses is concerned, any shape is theoretically possible so long as a stable configuration is adopted. The most favoured shapes are the traditional pitched truss and the parallel-chord truss (Fig. 6.28). The mansard shape is a commonly used variation on the latter. Bowstring shapes are also used but these are more expensive to construct (especially if they are at the higher end of the span range). Typical span ranges for all of these types are given in Table 6.9.

The internal geometry of a truss must be stable and this is achieved by adopting a fully triangulated arrangement. Where, for any reason, these types of geometry cannot be adopted, stability can be provided by continuity of elements through joints (Fig. 6.29). For efficiency in the use of material, internal angles of around 60° are best and angles which are less than 30° are avoided. Economy in construction results if the total number of joints is minimised, but this must not be achieved at the expense of ill-conditioned triangles (i.e. triangles with angles less than 30°) or the creation of excessively long compression elements. Adoption of a small number of joints also reduces the deflection which results from slipping at the fastenings.

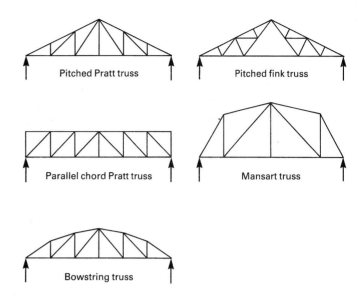

Fig. 6.28 Typical configurations for large one-off timber trusses.

Such slipping occurs with all types of mechanical fastener and can increase the total deflection of the truss significantly.

The overall depth of a timber truss must be such as to give a fairly small ratio of span to depth (Table 6.9) and this is particularly important where the span is long (greater than

Table 6.9 Typical span ranges and proportions for one-off timber trusses

Configuration		Span range (m)	Span/depth
Pitched		10–30	5
Parallel chord		15–45	8–10
Bowstring		15–75	6–8

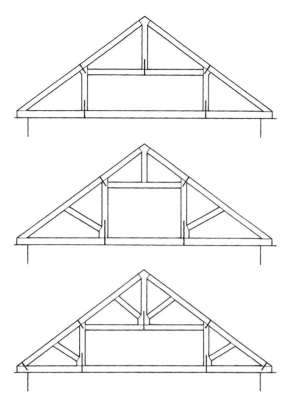

Fig. 6.29 Traditional forms of semi-truss such as these are not fully triangulated and depend on continuity of the sub-elements through the joints for stability.

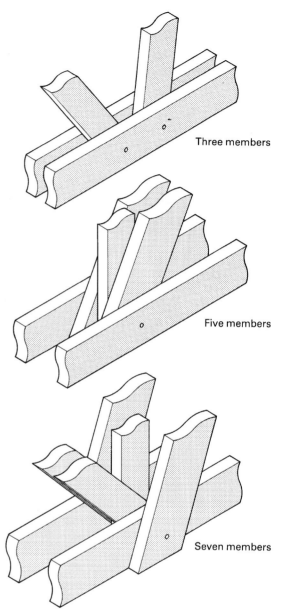

Three members

Five members

Seven members

Fig. 6.30 Typical joint arrangements for large, one-off trusses. The fastening elements are bolts which are used in conjunction with timber connectors.

20 m). Timber trusses must normally be significantly deeper than equivalent steel trusses and provision must be made for this in the design of buildings which are to have a primary structure of timber.

Most one-off trusses are assemblies of multiple-component elements which are connected together at multiple lap-joints in which the fastening components are bolts, usually with timber connectors (Figs 6.30 and 6.31). The critical factor which determines the feasibility or otherwise of a proposed arrangement is frequently the design of the joints (Fig. 6.32). This must be such as to allow sufficient area of overlap between elements to accommodate the number of connectors which are required to transmit the load. It is frequently found advantageous to use low-grade timber for large trusses because the larger elements which result from this provide more space for the joints. This also reduces the slenderness ratios of the compression elements.

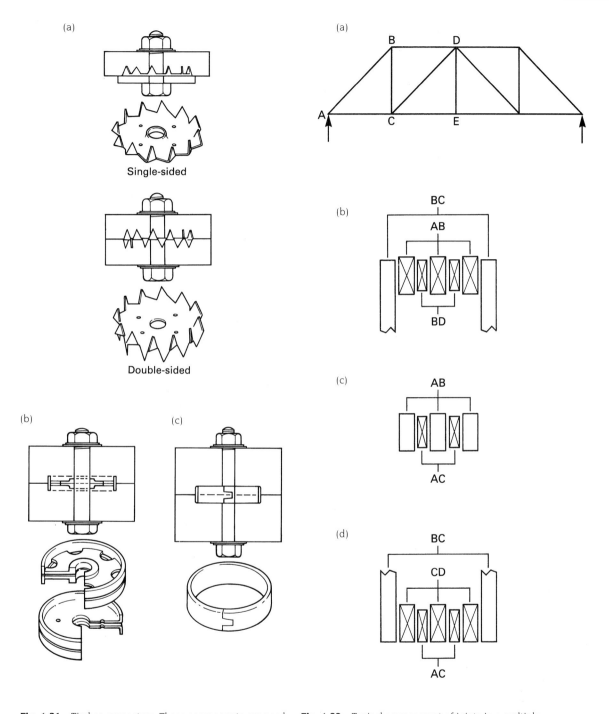

Fig. 6.31 Timber connectors. These components are used in conjunction with bolts to make lap-type connections between timber elements. Their function is to reduce the stress concentration which occurs at the bolt.
(a) Toothed-plate connector.
(b) Shear-plate connector.
(c) Split-ring connector.

Fig. 6.32 Typical arrangement of joints in a multiple-element large timber truss.
(a) Configuration of truss.
(b) Section through joint B.
(c) Section through joint A.
(d) Section through joint C.

Fig. 6.33 Detail of large one-off timber truss.

6.6.6 Joints

6.6.6.1 Introduction
Joints in structural timber are made either by direct bearing of one element on another, as in the case of joists bearing on a wall-plate or on joist hangers (Fig. 6.34) or with the aid of fastening components which transmit the load between elements in lap-type arrangements (Fig. 6.35). The design of the direct-bearing type of joint is straightforward and is rarely critical in determining the feasibility of a structure. Lap-type joints between elements in trusses, skeleton frames or built-up-beam

(a)

(b)

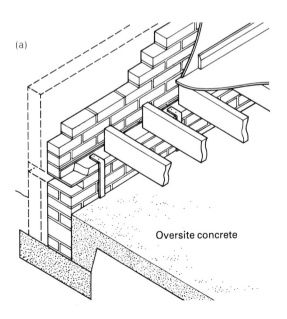

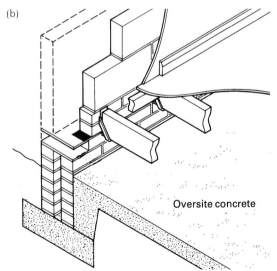

Oversite concrete

Oversite concrete

Fig. 6.34 Timber joists are normally supported by direct bearing, either on a wall-plate, as in (a) or on joist hangers, as in (b).

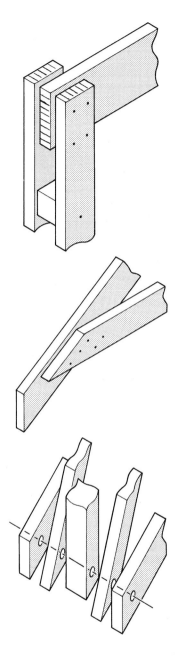

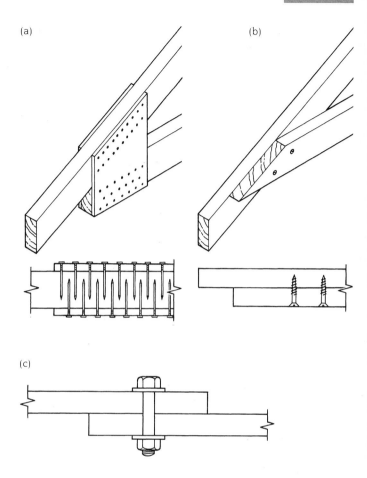

(a) (b)

(c)

Fig. 6.36 The three principal types of fastening component which are used in timber engineering are nails, screws and bolts.

Fig. 6.35 Lap-type joints. In each case the load is transmitted between elements by shear in the fastening components, which are either nails, screws or bolts.

components, can affect the feasibility of a proposed structure and are therefore an important consideration in the early stages of a design.

Two types of fastening medium are used for timber joints: mechanical fasteners (nails, screws, bolts, etc. (Fig. 6.36)) and adhesives. Additional jointing components such as gusset plates of plywood or steel are also frequently necessary and joints are normally designed so that the fastening medium is loaded in shear. A fairly wide choice of different possible arrangements will usually be available for a particular joint due to the range of fastening components which are available and to the possibility of placing elements either in a single plane and joining them through gusset plates, or making them overlap and joining them directly together in lap-type joints.

211

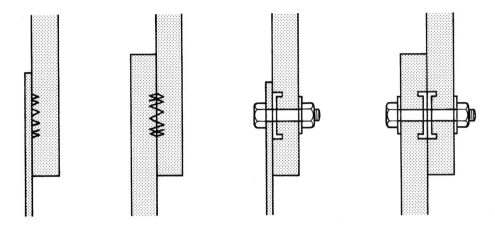

Fig. 6.37 Toothed-plate and shear-plate connectors are available in single- and double-sided formats and can be used either for timber-to-timber or timber-to-steel junctions.

The selection of the joint type is usually based on a number of considerations, such as economy (both in the cost of the fastening components and in the manufacture of the joint), the type of workshop facilities which will be available, the appearance of the joint, and any special requirements which are imposed by the way in which the structure will be assembled. A fundamental decision is whether to use mechanical fasteners or adhesive. Glued joints produce more satisfactory structural connections but are difficult to make on site. Components which are fastened with glued joints must normally therefore be produced under factory conditions and considerations of transport dictate the maximum size of these. Very large elements must be assembled on site and this will normally mean that mechanical fasteners will have to be used.

A critical factor in the design of timber joints, especially mechanical joints, is the weakness of the material in crushing and shear; the basic strategy which is adopted to overcome this is to reduce the concentration of stress which occurs by bringing the largest possible area of timber into play. In the case of a glued joint this is achieved by providing a large contact area on which the glue can act. Where mechanical fasteners are used it can be achieved by using a large number of small fastening elements (Fig. 6.36a) (although there are limits to the number of fasteners which can be placed in a single line due to the unevenness in load distribution which tends to develop between them) or by using individual components which are specifically designed to reduce concentrations of stress, such as timber connectors (Fig. 6.31). The required overall width of a timber element is frequently determined by the need to provide sufficient space to accommodate the fastening components which are required at the joint rather than by the need to provide a cross-section which is large enough to resist the internal force which the element carries.

6.6.6.2 Joints with mechanical fasteners
The principal types of mechanical fastener are nails, screws, bolts, timber connectors and punched metal plates. Nails and screws, which have similar dimensions, are capable of carrying similar amounts of load. Nails are normally preferred due to their lower primary cost and to the lower labour cost involved in driving them into the timber. Screws are used only where high resistance to withdrawal is required. Individual bolts carry higher loads than nails but the safe working load for a bolt is usually considerably lower than the strength of the bolt itself, due to the inability of the timber to accommodate the high concentration of stress which occurs at a bolted connection. Thus, although fewer bolts than nails may be required for a particular connection the total

Fig. 6.38 Punched-metal-plate fasteners, which form combined nail-and-gusset plate units, are used as the fastening elements in these trussed rafters.

size of the connection may be greater when bolts are used due to the larger spacing requirements combined with the relatively low load-carrying capacity of individual bolts in timber.

Timber connectors are the mechanical fasteners which can carry the highest loads and there are three types of these: toothed-plate connectors, shear-plate connectors and split-ring connectors (Fig. 6.31). Toothed-plate and shear-plate connectors are available in single or double-sided versions. The former are used for timber-to-plywood and timber-to-steel joints and the latter for timber-to-timber joints (Fig. 6.37). All timber connectors are used in conjunction with bolts, which carry a proportion of the load which passes through the joint. The principal function of the connector is to reduce the concentration of stress.

The detailed design of a mechanical joint which is made with nails, bolts or timber connectors involves decisions on the type of fastener to be used, the number which are required to provide the necessary strength, and the spacing which must be adopted between these. The choice of fastener type is affected by constructional factors and considerations of strength. Normally, nails and bolts (without connectors) are used for structures of modest size and timber connectors for the larger struc-

tures such as one-off trusses. The number of individual fasteners which are required for a joint depends on the load-carrying capacity of the fastener chosen in relation to the load which passes through the joint. This depends in turn on the size of the individual fastener, the species of the timber, and the dimensions of the timber. Individual fasteners must then be positioned so as to satisfy the minimum spacing requirements and in some cases it is necessary to increase the size of the element which is used in order to accommodate the fasteners safely.

Punched metal plates (Fig. 6.38) combine the functions of nails and gusset plates and are used to make connections in trussed rafters and trusses in which all sub-elements are in a single plane. They are not suitable for the largest trusses, however, and perform best in elements of relatively short span (up to 12 m). A range of products of this type is available and load values for design must be obtained direct from manufacturers. Calculations are rarely performed for this type of joint. Punched-metal-plate fasteners are not suitable for joints which must be assembled on site.

6.6.6.3 Glued joints
A range of adhesives which are suitable for making structural joints in timber and which are

213

resistant to a variety of environmental conditions is available. Both timber-to-timber connections and joints between timber and other materials can be made with glue. Glued joints are designed so that the glue is loaded in shear only and tests have indicated that when this type of joint is loaded to breaking point, the failure occurs in the timber adjacent to the glued surface and not in the glue itself. The allowable stresses on which the design of glued joints are based are therefore the shear stresses, parallel to the grain of the timber concerned; they are subject to the normal modification factors to allow for the effects of duration of load and the direction of load in relation to the grain. Although the rigorous design of a glued joint is a complicated process, the basic requirement is to provide an area of contact between the components which is large enough to maintain the shear stress at an acceptable level. The feasibility of a proposed joint, and therefore of the structural element of which it forms part, is frequently determined by the required size of the contact area. This must

obviously not be excessive in relation to the sizes of the sub-elements being joined.

An additional factor which must be considered in relation to the feasibility of specifying a glued joint is the condition of the environment in which the joint will be made. This should ideally be that of the workshop or factory. Most glues have to be cured in conditions of close environmental control and surfaces which are to be glued must be carefully prepared to achieve a glue line which is as thin as possible. Also, the surface to which the glue is applied should be freshly machined (not more than 48 hours old); this is the case because the open ends of cut cells, whose presence is essential to achieve effective bond with glue, tend to close with time. The standards of both environmental control and the type of supervision which are necessary for the manufacture of reliable glued joints are frequently difficult to achieve on site, and it is normal practice to specify mechanical fasteners rather than glue for site-made joints.

Fig. 6.39 Typical timber floor deck with closely spaced joists supporting a floor consisting of boarding.

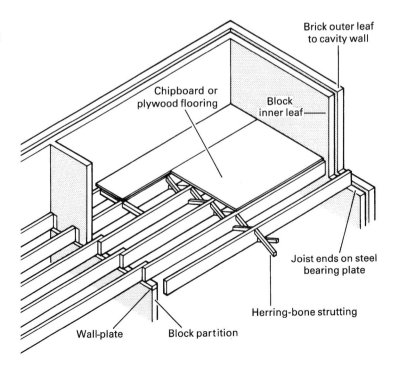

Brick outer leaf to cavity wall

Chipboard or plywood flooring

Block inner leaf

Joist ends on steel bearing plate

Herring-bone strutting

Wall-plate

Block partition

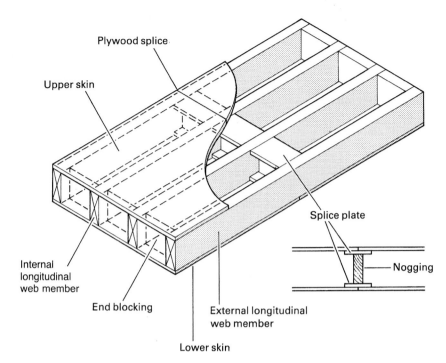

Plywood splice

Upper skin

Internal longitudinal web member

End blocking

External longitudinal web member

Lower skin

Splice plate

Nogging

6.7 Structural forms for timber

6.7.1 Floor and roof decks
Perhaps the simplest type of timber structure is the traditional floor or roof deck in which sawn-timber beams (joists) are spaced at close centres (450 mm to 600 mm) supporting a deck of relatively thin boarding (Fig. 6.39). Traditionally, the boarding was of machined solid timber boards which had tongued-and-grooved edges so as to spread the effect of concentrated load. In recent years plywood, blockboard and even chipboard have been used in this role. These provide the deck with greater resistance to in-plane loads than tongued-and-grooved forms of boarding, and give it the ability to act more effectively as diaphragm bracing, and therefore to contribute to the general stability of the structure.

Timber decks are normally used in conjunction with loadbearing-wall vertical elements (either of timber or masonry) but can also be used with skeleton frames. The range of spans for which the traditional joisted deck is normally used is 3.5 m to 6 m. Tables 6.10 and 6.11 give the spans which are possible for the standard range of sawn-timber sizes.

Two comparatively recent developments in this form of construction are composite decks, in which the joists and sheeting act together in T- or box-beam configurations, and the use of built-up-beams in place of sawn timber to form the joists. If plywood is used to form a floor or roof surface and is nailed and glued to the joists, a T- or box-beam structure is created which is stronger, and can span further, than a traditional boarded deck of equivalent dimensions (Fig. 6.40). The resulting element is called a 'stressed-skin' deck and this form of structure allows spans of up to 9 m to be achieved with the largest sizes of sawn-timber joist. Its use therefore extends the span range of flat timber decks made with sawn-timber elements. A good structural connection between the web members (internal stringers) and the skin is required if effective composite action is to develop, however, and the joint should preferably be made by dense nailing and gluing. The assembly of the deck is best carried out under factory conditions and

215

Table 6.10 Span ranges for sawn-timber joists in domestic floors (section sizes under the line are not in common use)

Span (mm)	300	350	400	450	500	550	600
				Spacing (mm)			
1200	38 × 75	44 × 75	50 × 75	38 × 100	38 × 100	38 × 100	44 × 100
1300	44 × 75	50 × 75	38 × 100	38 × 100	38 × 100	44 × 100	50 × 100
1400	44 × 75	50 × 75	38 × 100	38 × 100	44 × 100	50 × 100	50 × 100
1500	50 × 75	38 × 100	38 × 100	44 × 100	44 × 100	50 × 100	38 × 125
1600	38 × 100	38 × 100	38 × 100	44 × 100	50 × 100	38 × 125	38 × 125
1700	38 × 100	38 × 100	44 × 100	50 × 100	38 × 125	38 × 125	44 × 125
1800	38 × 100	38 × 100	44 × 100	50 × 100	38 × 125	44 × 125	44 × 125
1900	38 × 100	44 × 100	50 × 100	38 × 125	38 × 125	44 × 125	50 × 125
2000	38 × 100	44 × 100	50 × 100	38 × 125	44 × 125	44 × 125	50 × 125
2100	44 × 100	50 × 100	38 × 125	38 × 125	44 × 125	50 × 125	38 × 150
2200	44 × 100	50 × 100	38 × 125	44 × 125	50 × 125	50 × 125	38 × 150
2300	50 × 100	38 × 125	38 × 125	44 × 125	50 × 125	38 × 150	44 × 150
2400	38 × 125	38 × 125	44 × 125	50 × 125	50 × 125	38 × 150	44 × 150
2500	38 × 125	38 × 125	44 × 125	50 × 125	38 × 150	44 × 150	50 × 150
2600	38 × 125	44 × 125	50 × 125	38 × 150	44 × 150	50 × 150	50 × 150
2700	38 × 125	44 × 125	38 × 150	44 × 150	44 × 150	50 × 150	44 × 175
2800	44 × 125	50 × 125	38 × 150	44 × 150	50 × 150	38 × 175	44 × 175
2900	50 × 125	38 × 150	44 × 150	50 × 150	38 × 175	44 × 175	50 × 175
3000	38 × 150	38 × 150	44 × 150	50 × 150	44 × 175	44 × 175	50 × 175
3100	38 × 150	44 × 150	50 × 150	44 × 175	44 × 175	50 × 175	44 × 200
3200	38 × 150	44 × 150	50 × 150	44 × 175	50 × 175	50 × 175	44 × 200
3300	44 × 150	50 × 150	44 × 175	44 × 175	50 × 175	44 × 200	50 × 200
3400	44 × 150	38 × 175	44 × 175	50 × 175	44 × 200	44 × 200	50 × 200
3500	50 × 150	44 × 175	44 × 175	50 × 175	44 × 200	50 × 200	50 × 200
3600	38 × 175	44 × 175	50 × 175	44 × 200	44 × 200	50 × 200	44 × 255
3700	38 × 175	44 × 175	50 × 175	44 × 200	50 × 200	44 × 225	50 × 225
3800	44 × 175	50 × 175	44 × 200	50 × 200	50 × 200	44 × 225	50 × 225
3900	44 × 175	50 × 175	44 × 200	50 × 200	44 × 225	50 × 225	50 × 225
4000	50 × 175	44 × 200	44 × 200	50 × 200	44 × 225	50 × 225	75 × 200
4100	50 × 175	44 × 200	50 × 200	44 × 225	50 × 225	50 × 225	75 × 200
4200	44 × 200	44 × 200	50 × 200	44 × 225	50 × 225	75 × 200	75 × 200
4300	44 × 200	50 × 200	44 × 225	50 × 225	75 × 200	75 × 200	50 × 250
4400	44 × 200	50 × 200	44 × 225	50 × 225	75 × 200	75 × 200	75 × 225
4500	44 × 200	44 × 225	50 × 225	50 × 225	75 × 200	50 × 250	75 × 225
4600	25 × 200	44 × 225	50 × 225	75 × 250	50 × 250	75 × 225	75 × 225
4700	50 × 200	50 × 225	50 × 225	75 × 250	50 × 200	75 × 225	75 × 225
4800	44 × 225	50 × 225	75 × 200	50 × 250	75 × 225	75 × 225	75 × 250
4900	44 × 225	50 × 225	50 × 250	50 × 250	75 × 225	75 × 225	75 × 250
5000	44 × 225	50 × 225	50 × 250	75 × 250	75 × 225	75 × 225	75 × 250
5100	50 × 225	50 × 225	50 × 250	75 × 250	75 × 225	75 × 225	75 × 250
5200	50 × 225	50 × 250	50 × 250	75 × 225	75 × 250	75 × 250	75 × 250
5300	75 × 200	50 × 250	75 × 225	75 × 250	75 × 250	75 × 250	75 × 300
5400	50 × 250	50 × 250	75 × 225	75 × 250	75 × 250	75 × 250	75 × 300
5500	50 × 250	50 × 250	75 × 250	75 × 250	75 × 250	75 × 300	75 × 300
5600	50 × 250	75 × 225	75 × 250	75 × 250	75 × 250	75 × 300	75 × 300
5700	50 × 250	75 × 250	75 × 250	75 × 250	75 × 250	75 × 300	75 × 300
5800	50 × 250	75 × 250	75 × 250	75 × 250	75 × 300	75 × 300	75 × 300
5900	75 × 225	75 × 250	75 × 250	75 × 300	78 × 300	78 × 300	75 × 300
6000	75 × 250	75 × 250	75 × 250	75 × 300	75 × 300	75 × 300	75 × 300
6100	75 × 250	75 × 250	75 × 300	75 × 300	75 × 300	75 × 300	75 × 300
6200	75 × 250	75 × 250	75 × 300	75 × 300	75 × 300	75 × 300	75 × 300
6300	75 × 250	75 × 300	75 × 300	75 × 300	75 × 300	75 × 300	
6400	75 × 250	75 × 300	75 × 300	75 × 300	75 × 300	75 × 300	
6500	75 × 250	75 × 300	75 × 300	75 × 300	75 × 300	75 × 300	
6600	75 × 300	75 × 300	75 × 300	75 × 300	75 × 300		
6700	75 × 300	75 × 300	75 × 300	75 × 300	75 × 300		
6800	75 × 300	75 × 300	75 × 300	75 × 300			
6900	75 × 300	75 × 300	75 × 300	75 × 300			
7000	75 × 300	75 × 300	75 × 300				
7100	75 × 300	75 × 300	75 × 300				
7200	75 × 300	75 × 300	75 × 300				
7300	75 × 300	75 × 300					
7400	75 × 300	75 × 300					
7500	75 × 300						
7600	75 × 300						
7700	75 × 300						
7800	75 × 300						

Table 6.11 Span ranges for sawn-timber joists in flat roof decks to which pedestrians do not have access (section sizes under the line are not in common use)

Span (mm)	Spacing (mm) 300	350	400	450	500	550	600
1200	38 × 75	38 × 75	38 × 75	38 × 75	38 × 75	38 × 75	38 × 75
1300	38 × 75	38 × 75	38 × 75	38 × 75	38 × 75	38 × 75	38 × 75
1400	38 × 75	38 × 75	38 × 75	38 × 75	38 × 75	38 × 75	38 × 75
1500	38 × 75	38 × 75	38 × 75	38 × 75	38 × 75	44 × 75	44 × 75
1600	38 × 75	38 × 75	38 × 75	38 × 75	38 × 75	38 × 75	38 × 75
1700	38 × 75	38 × 75	38 × 75	44 × 75	50 × 75	38 × 100	38 × 100
1800	38 × 75	38 × 75	44 × 75	50 × 75	38 × 100	38 × 100	38 × 100
1900	38 × 75	44 × 75	50 × 75	38 × 100	38 × 100	38 × 100	38 × 100
2000	44 × 75	50 × 75	38 × 100	38 × 100	38 × 100	38 × 100	38 × 100
2100	50 × 75	38 × 100	38 × 100	38 × 100	38 × 100	38 × 100	44 × 100
2200	38 × 100	38 × 100	38 × 100	38 × 100	38 × 100	44 × 100	50 × 100
2300	38 × 100	38 × 100	38 × 100	38 × 100	44 × 100	50 × 100	50 × 100
2400	38 × 100	38 × 100	38 × 100	44 × 100	50 × 100	38 × 125	38 × 125
2500	38 × 100	38 × 100	44 × 100	50 × 100	38 × 125	38 × 125	38 × 125
2600	38 × 100	44 × 100	50 × 100	38 × 125	38 × 125	38 × 125	38 × 125
2700	44 × 100	50 × 100	38 × 125	38 × 125	38 × 125	38 × 125	44 × 125
2800	50 × 100	38 × 125	38 × 125	38 × 125	38 × 125	44 × 125	50 × 125
2900	50 × 100	38 × 125	38 × 125	38 × 125	44 × 125	50 × 125	50 × 125
3000	38 × 125	38 × 125	38 × 125	44 × 125	50 × 125	38 × 150	38 × 150
3100	38 × 125	38 × 125	44 × 125	50 × 125	38 × 150	38 × 150	38 × 150
3200	38 × 125	44 × 125	50 × 125	38 × 150	38 × 150	38 × 150	44 × 150
3300	38 × 125	44 × 125	50 × 125	38 × 150	38 × 150	44 × 150	44 × 150
3400	44 × 125	50 × 125	38 × 150	38 × 150	44 × 150	44 × 150	50 × 150
3500	50 × 125	38 × 150	38 × 150	44 × 150	44 × 150	50 × 150	38 × 175
3600	50 × 125	38 × 150	38 × 150	44 × 150	50 × 150	38 × 175	38 × 175
3700	38 × 150	38 × 150	44 × 150	50 × 150	38 × 175	38 × 175	38 × 175
3800	38 × 150	44 × 150	44 × 150	50 × 150	38 × 175	38 × 175	38 × 175
3900	38 × 150	44 × 150	50 × 150	38 × 175	38 × 175	44 × 175	50 × 175
4000	44 × 150	50 × 150	38 × 175	38 × 175	44 × 175	50 × 175	50 × 175
4100	44 × 150	50 × 150	38 × 175	44 × 175	44 × 175	50 × 175	44 × 200
4200	50 × 150	38 × 175	38 × 175	44 × 175	50 × 175	44 × 200	44 × 200
4300	50 × 150	38 × 175	44 × 175	50 × 175	44 × 175	44 × 200	44 × 200
4400	38 × 175	38 × 175	44 × 175	50 × 175	44 × 200	44 × 200	44 × 200
4500	38 × 175	44 × 175	50 × 175	44 × 200	44 × 200	44 × 200	50 × 200
4600	38 × 175	44 × 175	44 × 200	44 × 200	44 × 200	50 × 200	50 × 200
4700	44 × 175	50 × 175	44 × 200	44 × 200	50 × 200	50 × 200	44 × 225
4800	44 × 175	44 × 200	44 × 200	44 × 200	50 × 200	44 × 225	44 × 225
4900	50 × 175	44 × 200	44 × 200	50 × 200	44 × 225	44 × 225	44 × 225
5000	50 × 175	44 × 200	44 × 200	50 × 200	44 × 225	44 × 225	50 × 225
5100	44 × 200	44 × 200	50 × 200	44 × 225	44 × 225	44 × 225	50 × 225
5200	44 × 200	44 × 200	50 × 200	44 × 225	44 × 225	50 × 225	75 × 200
5300	44 × 200	50 × 200	44 × 225	44 × 225	50 × 225	50 × 225	50 × 250
5400	44 × 200	50 × 200	44 × 225	44 × 225	50 × 225	50 × 250	50 × 250
5500	50 × 200	44 × 225	44 × 225	50 × 225	75 × 200	50 × 250	50 × 250
5600	50 × 200	44 × 225	44 × 225	50 × 225	50 × 250	50 × 250	50 × 250
5700	44 × 225	44 × 225	50 × 225	50 × 250	50 × 250	50 × 250	50 × 250
5800	44 × 225	44 × 225	50 × 225	50 × 250	50 × 250	50 × 250	75 × 225
5900	44 × 225	50 × 225	50 × 250	50 × 250	50 × 250	75 × 225	75 × 250
6000	44 × 225	50 × 225	50 × 250	50 × 250	50 × 250	75 × 225	75 × 250
6100	44 × 225	75 × 200	50 × 250	50 × 250	75 × 225	75 × 250	75 × 250
6200	50 × 225	50 × 250	50 × 250	50 × 250	75 × 250	75 × 250	75 × 250
6300	50 × 225	50 × 250	50 × 250	75 × 225	75 × 250	75 × 250	75 × 250
6400	50 × 250	50 × 250	50 × 250	75 × 250	75 × 250	75 × 250	75 × 250
6500	50 × 250	50 × 250	75 × 225	75 × 250	75 × 250	75 × 250	75 × 300
6600	50 × 250	50 × 250	75 × 250	75 × 250	75 × 250	75 × 250	75 × 300
6700	50 × 250	50 × 250	75 × 250	75 × 250	75 × 250	75 × 300	75 × 300
6800	50 × 250	75 × 225	75 × 250	75 × 250	75 × 250	75 × 300	75 × 300
6900	50 × 250	75 × 250	75 × 250	75 × 250	75 × 300	75 × 300	75 × 300
7000	50 × 250	75 × 250	75 × 250	75 × 250	75 × 250	75 × 300	75 × 300
7100	75 × 225	75 × 250	75 × 250	75 × 300	75 × 300	75 × 300	75 × 300
7200	75 × 250	75 × 250	75 × 250	75 × 300	75 × 300	75 × 300	75 × 300
7500	75 × 250	75 × 250	75 × 300	75 × 300	75 × 300	75 × 300	75 × 300
7700	75 × 250	75 × 300	75 × 300	75 × 300	75 × 300	75 × 300	75 × 300
7800	75 × 250	75 × 300	75 × 300	75 × 300	75 × 300	75 × 300	
8000	75 × 300	75 × 300	75 × 300	75 × 300	75 × 300		
8200	75 × 300	75 × 300	75 × 300	75 × 300			
8400	75 × 300	75 × 300	75 × 300				
8500	75 × 300	75 × 300	75 × 300				
8600	75 × 300	75 × 300	75 × 300				
8800	75 × 300	75 × 300					
9000	75 × 300						
9200	75 × 300						
9300	75 × 300						

Table 6.12 Simple rules for preliminary sizing of stressed-skin panels

Parameter	Imposed load	
	0.75 kN/m²	1.5 kN/m²
Minimum overall depth	$\frac{span}{40}$	$\frac{span}{35}$
Minimum stringer width	50 mm	50 mm
Maximum stringer spacing	600 mm	400 mm
Tension skin	8 mm	9 mm
Compression skin	9 mm or 12 mm	12 mm, 15 mm or 18 mm

stressed-skin panels are normally regarded as components which must be prefabricated rather than made on site. They are used where longer spans are required than are possible with traditional forms of timber deck or where a smaller constructional depth must be provided. Data for stressed-skin panels are given in Table 6.12.

The second development of the traditional deck is the use of built-up-beams in place of sawn-timber joists for the principal elements (Fig. 6.41). Plyweb and laminated timber beams, as well as various proprietary lattice-beam types have been used in this role and these allow much larger spans (up to 20 m) to be achieved than are possible with sawn-timber elements. Economy in the use of these more sophisticated components requires that they be spaced further apart than traditional joists and beam spacings of around 1.5 m to 3.5 m are usually adopted. This requires that a fairly thick form of boarding be used to form the deck skin.

6.7.2 Pitched roof structures

In the traditional pitched roof structure, sawn-timber rafters are positioned at close centres (450 mm to 600 mm) supporting a roof skin of sarking boards or tile battens (Fig. 6.42). The vertical support structure is usually of the loadbearing-wall type and, because the rafters produce an outwards thrust at the wallhead, the system is suitable for small spans only (up to 3.5 m). The span range can be extended by use of tie elements, and in traditional construction various arrangements of semi-trussing were devised to allow longer spans to be achieved (Fig. 6.43). Traditional roof forms of this type require fairly large sizes of timber, however, because the principal elements are stressed in bending as well as axially, and so must have large cross-sections to resist this. Also, many of their geometries are fundamentally unstable and depend on the continuity of elements through joints for stability. Elements

Fig. 6.41 Plyweb built-up-beams at close spacing are used here to form the principal structural elements of a roof deck. The arrangement is similar to that of a traditional joisted deck with sawn-timber elements. The use of built-up-beams allows longer spans to be achieved.

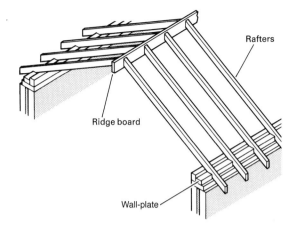

Fig. 6.42 Traditional pitched roof arrangement of closely spaced rafters and ridge board. The rafters exert a horizontal load on the tops of the walls.

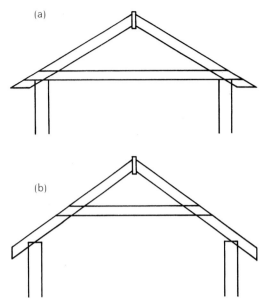

must therefore be long as well as of large cross-section. This method of construction has now been largely superseded by the trussed-rafter system in which rafters are incorporated into a fully triangulated arrangement (Fig. 6.27). Very small timber sections may be employed with this arrangement and savings are possible of up to 30% in the volume of timber which is used, as compared with traditional methods. Trussed rafters can be built up on site using bolts or nails but more commonly are proprietary components jointed with punched-metal-plate fasteners. They are most suited to spans in the range 5 m to 11 m. A disadvantage of the trussed-rafter roof system is that the structure occupies the whole of the roof space.

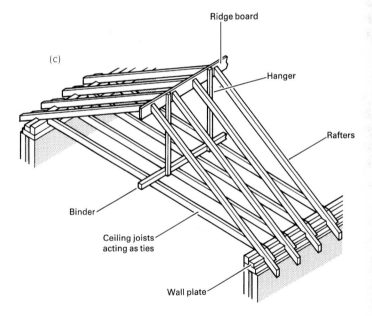

Fig. 6.43 Semi-trussed arrangements reduce or eliminate the horizontal force on the wallhead but require long sub-elements.
(a) The tie at eaves level eliminates horizontal thrust at the wallhead. The tie is here spanning the entire width of the building, which limits the span.
(b) The collared roof leaves some horizontal thrust at eaves level but requires a shorter length of tie.
(c) A longer span is possible if the tie is supported from the ridge.
(d) This semi-trussed arrangement is dependent on continuity of elements through joints for stability.

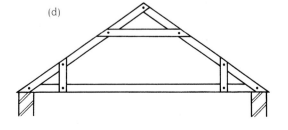

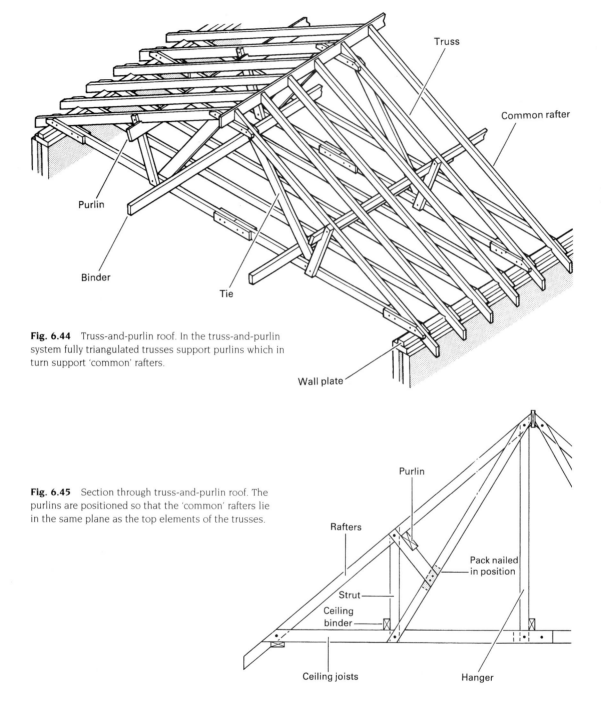

Fig. 6.44 Truss-and-purlin roof. In the truss-and-purlin system fully triangulated trusses support purlins which in turn support 'common' rafters.

Fig. 6.45 Section through truss-and-purlin roof. The purlins are positioned so that the 'common' rafters lie in the same plane as the top elements of the trusses.

An alternative to trussed-rafter construction is purlin construction in which plain rafters are supported on a series of secondary beams, called purlins, which run parallel to the direction of the ridge. The purlins can be supported on trusses or cross-walls of timber-stud or masonry (Figs 6.44 to 6.47). Where trusses are used they are usually spaced 2 m to 3 m apart.

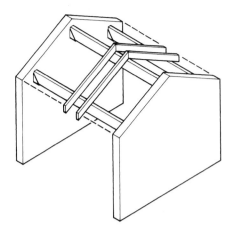

Fig. 6.46 Purlin roof supported on cross-walls. This arrangement leaves the roof space free of structure. The purlins carry a relatively large area of roof, however, and therefore significant load. Built-up-beam sections are required to achieve spans greater than 3 m.

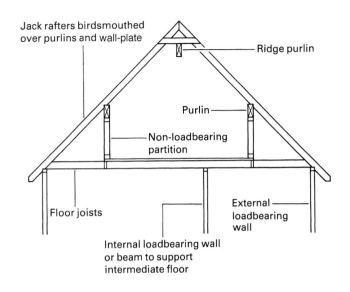

Fig. 6.47 Typical cross-section through a purlin roof.

Cross-wall spacing is normally wider (5 m to 6 m). The trusses which are used in truss-and-purlin construction are much stronger and heavier than trussed rafters and usually have sub-elements which consist of several boards jointed by timber connectors. Alternatively, a truss can be made by bolting together a number of trussed rafters.

Purlin construction has the advantage of leaving the roof space relatively clear of structure. It also allows the longitudinal walls of a building to be non-loadbearing, which can be an advantage with certain types of planning. The purlins themselves tend to be fairly heavily loaded, however, and must normally be timbers of large cross-section. The size of cross-section which is required depends on the purlin spacing and on the weight of the roof which is carried, but the largest sizes of sawn-timber elements are rarely capable of spanning further than around 2.5 m when acting as purlins. Where larger spans are required, built-up timber beams or lightweight steel elements are used. The depth of built-up timber purlins is likely to be large (800 mm for spans in the region of 5 m to 6 m and provision for this must be made in the planning of the building.

In truss-and-purlin structures it is common practice for the top element of the truss to be located in the same plane as the rafters as this simplifies the detailing of the roof. The arrangement requires that purlins be suspended from the top elements of the trusses rather than located on top of them (Fig. 6.45) as was the normal practice in traditional roofs (see Section 6.2). Span ranges for the types of timber roof described here are given in Table 6.13.

Table 6.13 Approximate span ranges for different types of timber pitched-roof structures

		Span
Spanning across building	Timber joist (flat), plain rafter, close couple	2.5 m to 6 m
	Tied rafter	3 m to 7 m
	Trussed rafter	6 m to 20 m
	Truss and purlin	10 m to 30 m
Spanning between cross-walls	Sawn-timber purlin	2.5 m to 3.5 m
	Plyweb, laminated or steel purlin	2.5 m to 8 m

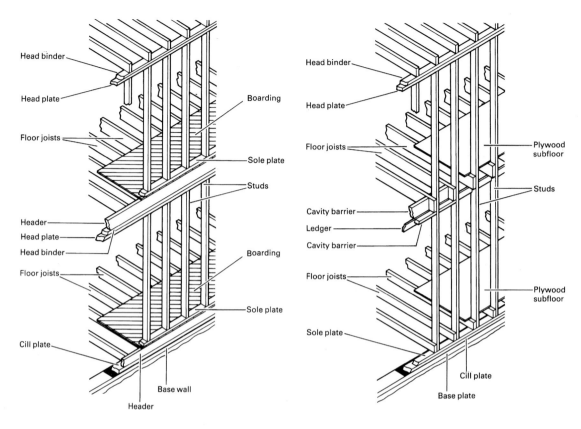

Fig. 6.48 Timber wallframe – platform frame type. In the platform frame the individual wall and floor panels are independent of each other. The sequence of construction is that the ground-floor walls are erected first, followed by the floors. These form a 'platform' on which the first-floor walls are then erected. The system lends itself to prefabrication and has the additional advantage that none of the individual elements is of great length. Its disadvantage is that the level of structural integrity is lower than with other types of arrangement.

Fig. 6.49 Timber wallframe – balloon frame type. In the balloon frame the studs which form the walls are continuous through two floors and the floor joists are attached directly to them. This arrangement provides better structural integrity than the platform frame and produces a weathertight enclosure more quickly. It cannot be prefabricated, however, and requires the use of very long timber elements.

6.7.3 Timber wallframe structures

Timber wallframes are a type of loadbearing-wall structure. They are sometimes called simply timber frame structures but they are not true skeleton frames. In wallframes, timber wall-panels (stud panels) consisting of elements of small cross-section placed at close centres are used to support traditional forms of roof and floor structure (Figs 6.48 to 6.51). The overall size of the structure is usually small: there are rarely more than two storeys and floor spans are in the range 3.5 m to 5 m with roof spans of around 6 m to 12 m. The system is used principally for domestic buildings.

The basic planning requirements for timber wallframe buildings are the same as for all loadbearing-wall structures. The wall plan, which must be more-or-less the same on every storey, must be capable of supporting all areas of roof and floor and, as both the floor and roof structures are one-way-spanning systems, this requires an arrangement of primary loadbearing walls which are parallel to each other and spaced at roughly equal distances apart. A typical arrangement is shown in

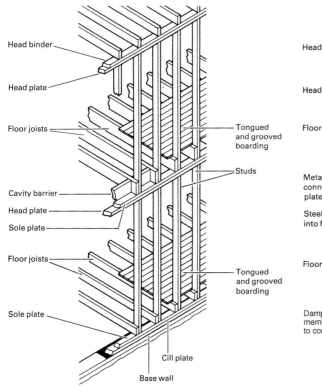

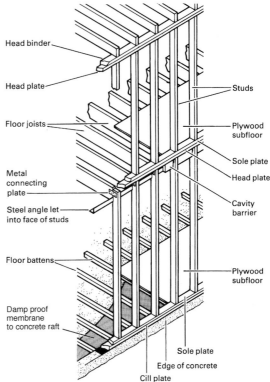

Fig. 6.50 Timber wallframe – modified frame type. In the modified frame the wall elements are only one storey high but the floor joists are attached directly to them. The modified frame is a compromise between the platform and balloon arrangements which combines some of the advantages of both.

Fig. 6.51 Timber wallframe – the independent frame type. In the independent frame the wall elements are one storey high but the floor joists are carried on a steel angle attached to the inner face of the wall. This arrangement lends itself to prefabrication. The fact that the floor joists do not penetrate the outer wall improves the thermal efficiency of the wall.

Fig. 6.52. Note that, as in the case of masonry structures, spine- and cross-wall forms are used. In spine-wall arrangements the roof normally spans the full width of the building and only the floors are supported on intermediate walls. Cross-walls are normally used in conjunction with purlin roof construction.

The wall-to-floor and wall-to-roof junctions in this form of structure are non-rigid and lateral stability is provided by the wall, floor and roof panels themselves, which serve as diaphragm bracing in the vertical and horizontal planes. Care must be taken in the planning of the buildings to ensure that vertical-plane bracing is correctly positioned. Bracing walls must be provided which are as symmetrically disposed as possible on plan.

It is also necessary that both the wall and the floor panels should have sufficient strength to resist in-plane racking[7] caused by wind load. This is done by sheathing with plywood or some other form of composite board

---

7 The tendency of the original rectangular shape to become a parallelogram due to shearing action.

First-floor plan

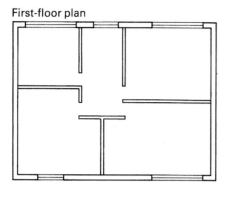

Ground-floor plan

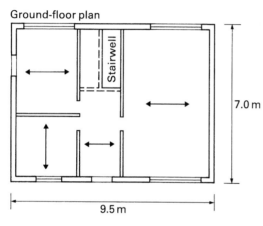

Stairwell

7.0 m

9.5 m

Fig. 6.52 Typical plan-form of a small-scale timber wallframe building. The arrangement conforms to the parallel-wall format of all loadbearing-wall forms of construction.

(traditional tongued-and-grooved boarding does not provide good racking resistance due to the possibility of individual boards slipping relative to each other) or by incorporating diagonal bracing (tensioned steel wire or timber) into the wall and floor panels.

A number of different wall configurations are used in timber loadbearing-wall structures. The traditional methods are the 'platform' frame or the 'balloon' frame. In platform frames (Fig. 6.48) the wall and floor panels are more-or-less independent entities and the wall panels are of single-storey height. Each wall is built on top of the platform which is formed by the floor below it and the floor panels therefore penetrate through the structural part of the wall to the inner side of the cladding. The wall

studs are positioned at the same spacing as the floor joists so as to minimise eccentricity in the transfer of load. The advantages of the platform frame are that long timber elements are not required and that the system lends itself to prefabrication. This is due to the relative independence of the floors from the walls and to the fact that none of the panels is very large and difficult to transport. The fact that studs and floor joists are aligned is also an advantage because it eases the constructional planning by allowing a grid system to be used. The principal disadvantages of the platform frame are that the building does not become a weathertight shell until the construction is at a fairly advanced stage, and that the structure is slightly less rigid than alternative forms.

In balloon frames (Fig. 6.49) the wallframes are two storeys in height (this is normally the full height of the building) and the floor joists are attached individually to wall studs by lap joints. The advantage of this system is that it allows the roof to be completed at an early stage in the construction process, before the floors are in position, to provide a weathertight shell. It also provides a structure which is more rigid than the platform frame. Its disadvantages are that it requires very long elements for the wall stud, that it does not lend itself to prefabrication and that the joists and studs are not aligned.

Two other types of wallframe are the 'modified' frame and the 'independent' frame (Figs 6.50 and 6.51). Both have wall panels which are one storey high. In the modified frame the ground- and first-floor wall panels are attached directly to each other and the floor joists are then attached directly to wall studs, as in the balloon frame. In the independent frame the same arrangement for wall panels is used but the floor is supported by a steel angle or other element attached to the edge of the lower panel and does not penetrate the thickness of the wall. The modified and independent frames are intended to combine the good features of the platform and balloon frames. They achieve something of the structural continuity of the latter without

Fig. 6.53 Single-storey timber skeleton-frame structure in which the principal elements are plyweb built-up-beam sections.

requiring very long elements of small cross-section, and they allow some degree of prefabrication to be possible.

6.7.4 The skeleton frame

The fact that timber possesses tensile, compressive and bending strength makes possible the construction of skeleton frames in the material (Fig. 6.53). A characteristic of this type of structure, however, is that individual elements are subjected to fairly large amounts of internal force, due to the concentration of the load into a small volume of structure, and

while this can be accommodated easily with strong materials, such as steel or reinforced concrete, it can be problematical with timber.

Where a skeleton frame is constructed in timber the whole of the structure will normally be rather bulky, especially if floor loads are carried. The structure must normally, therefore, be treated as one of the special architectural features of the building which it supports.

The floor- and roof-deck systems which are used are usually also of timber and of the one-way-spanning type. The frames are therefore best planned on a rectangular grid of beams

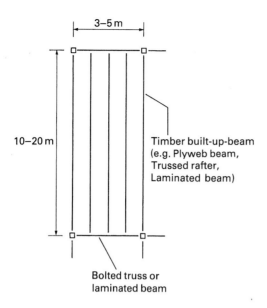

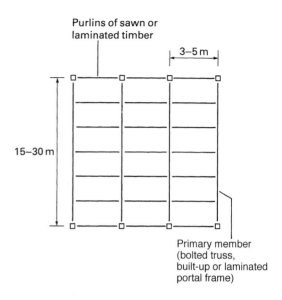

Fig. 6.54 Beam grid for single-storey timber skeleton frame. Closely spaced joists (built-up-beam sections) are carried on primary beams of relatively short span.

Fig. 6.55 Beam grid for single-storey timber skeleton frame. Strong primary structural elements (one-off trusses/ portal frameworks) carry a secondary structure of purlins.

and columns. In floor structures the floor joists should span parallel to the long side of the rectangle and the main beams parallel to the short sides.

The most successful timber skeleton frames are single-storey structures carrying low levels of imposed load (Fig. 6.53). Fairly long spans are possible with this type of arrangement, especially if built-up-beams or trusses are used for the principal elements, but the sizes of the elements are likely to be large compared to equivalents in steel.

Two strategies may be adopted for the planning of single-storey frames (Figs 6.54 and 6.55). In Fig. 6.54 the main structural elements span in the long direction in a rectangular column grid and are spaced close together to minimise the load carried by individual elements, and to allow them to carry the cladding directly without the use of a secondary structure. The geometry of this type of arrangement is similar to that of a steel frame with lightweight joists (Figs 3.25 and 3.27). Columns are positioned at every third or fourth beam and intermediate members are

carried on 'primary' beams. The latter are more heavily loaded than the main beams but can usually be of the same depth due to their smaller span. The exact configuration which is adopted in a particular case depends on the type of beam being used.

The second type of arrangement is one in which the primary beams are placed at a wider spacing and a secondary structure, on which the cladding is mounted, is provided to span between these (Fig. 6.55). The column grid coincides with the primary beams, which must be much stronger elements than those in the system described previously. Proprietary beams (at least in the simply supported form) are rarely suitable for this type of arrangement and large trusses are normally used. Built-up-beam-type cross-sections can be used as primary elements if a 'strong' structural configuration, such as a portal frame (Figs 6.3 and 6.53) is adopted. This type of structure usually has a plan grid which is similar to those which are used in single-storey steel frames (Fig. 3.30) but in the case of timber the primary elements are usually placed closer together

Table 6.14 Span range and dimensions of principal elements in timber trusses in single-storey skeleton frameworks

Span (m)	Truss spacing (m)	Roof pitch	Approximate member size (mm)	
			Rafter	Ceiling tie
10	2	20°	1 × 50 × 150S	1 × 50 × 100S
		30°	1 × 50 × 125S	1 × 50 × 75S
15	3	20°	1 × 75 × 200S	1 × 75 × 150S
		30°	1 × 75 × 150S	1 × 75 × 150S
20	4	20°	2 × 75 × 225S	1 × 75 × 225S
		30°	2 × 75 × 175S	1 × 75 × 150S
30	6	20°	4 × 75 × 200S	3 × 75 × 200S
		30°	3 × 75 × 125S	2 × 75 × 200S
40	6	20°	1 × 150 × 300L	Steel tie rod
		30°	1 × 150 × 250L	Steel tie rod
50	6	20°	1 × 150 × 330L	Steel tie rod
		30°	1 × 150 × 300L	Steel tie rod
60	6	20°	1 × 150 × 400L	Steel tie rod
		30°	1 × 150 × 330L	Steel tie rod

S = sawn timber
L = laminated timber

(3.5 m to 6 m). Element sizes for two versions of this type of frame are given in Table 6.14.

Where timber is used for multi-storey skeleton-frame structures and must carry floor loading, the spans must be kept small – around 2 m to 3 m if simple sawn-timber elements are used for floor beams. Longer spans are possible with built-up-beam elements, such as laminated timber or plyweb beams, but the depths of these will be very large compared to equivalent steel or reinforced concrete structures (Fig. 6.1).

In multi-storey frames very simple grids, with columns at all grid points, are adopted where solid timber elements are used, so as to minimise the loads which individual beams carry (Fig. 6.56). Spaced beams are frequently adopted to increase the strength of individual elements and this allows longer spans to be achieved. Where built-up-beams are used, more complex primary/secondary-beam arrangements are possible. In all cases the beam depths must be checked at an early stage in the design as these are likely to be critical.

Another feature which is likely to be critical in skeleton-frame design is the geometry of the

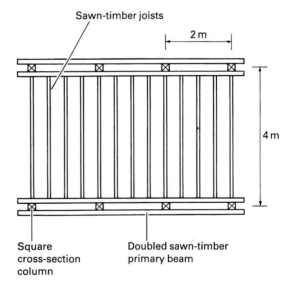

Sawn-timber joists

2 m

4 m

Square cross-section column

Doubled sawn-timber primary beam

Fig. 6.56 An example of a beam layout which is suitable for a multi-storey skeleton frame in timber. Primary elements are spaced beams (double elements) and are confined to short spans.

beam-to-column connections and some thought must normally be devoted to this when the design is at a preliminary stage. Simple lap joints with bolted connections are

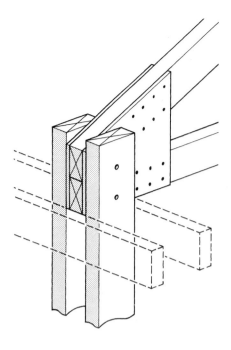

Fig. 6.57 Typical joint arrangements for timber skeleton frames. One of the features of this type of construction is the space required for joints.

normally used and, to avoid eccentricity in the transfer of load between elements, a 'spaced' arrangement can be adopted for either the beams or the columns (Fig. 6.57). In frames with solid timber beams the strength of the beams is usually a critical factor and so spaced beams are employed, with single-element columns. Where the horizontal elements are trusses or built-up-beams a spaced column can be used. Where tie-beams are required in the direction at right angles to that of the principal members, these will frequently have to be placed at a different level from the principal elements, so as to allow sufficient space for the two sets of lap joints. This increases the total depth of the structural zone and can be a critical factor in the design.

Beam-to-column joints in timber frames are rarely rigid and vertical-plane bracing is therefore required in skeleton frames. This can be provided either in the form of diagonal bracing (usually of the tensioned-wire type) or by suitably positioned wall panels which act as vertical-plane diaphragms. As with skeleton frames in other materials, the braced panels should be positioned as symmetrically as possible on plan and the disposition of the vertical-plane bracing is therefore a factor which affects the internal planning of the building.

6.7.5 Shells and other surface forms

Surface structures are those in which the elements which form the skin of the structure have an important, sometimes dominant, structural role. Stressed-skin panels (Fig 6.40) are in this category; folded forms and curved shell structures are other examples and these are described briefly here.

Folded forms are arrangements of flat structural panels; there are two basic types, the 'prismatic' type, which consists of flat-sided corrugations whose depth is constant across the structure (Fig. 6.58) and more complex types in which the depth of the corrugations is varied. The latter are normally based on a system of triangular panels. In both cases the panels which form the structures usually consist of a double skin of plywood sheets separated by solid timber stringers. Nailed and glued joints are used to assemble the panels, which are then bolted together to form the complete structure. The structures are highly efficient, in the sense that they achieve a high strength and stiffness for the amount of material which they contain, and they are therefore suitable for long-span structures (up to 40 m) or for situations where self-weight must be minimised. They are sophisticated forms, however, which are difficult to design and construct and they are not in common use.

Timber shells, which are curved stressed-skin structures of the form-active type, are also highly efficient structures which achieve high strength and stiffness for a given weight of material. Like folded forms they are difficult to construct and are not in common use. Unlike folded forms, timber shells are usually solid and constructed in layers. Strips of thin boarding are laid across a formwork or mould of the required shape and several layers, which are nailed and glued together, are used. These are

Fig. 6.58 Timber folded-plate struc- ture. Forms such as this, which are constructed from panels consisting of two skins of plywood separated by sawn- timber stringers, have good structural properties and allow long spans to be achieved.

Fig. 6.59 Timber hyperbolic paraboloid shells are used here to form a roof structure of distinctive shape.

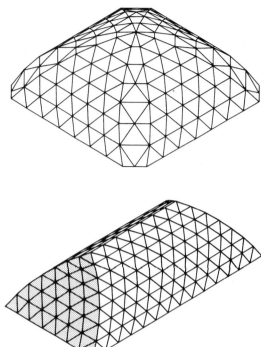

Fig. 6.60 Basic forms of the timber braced dome and lamella vault.

Fig. 6.61 The braced dome configuration allows large interiors to be constructed with a very efficient use of structural material.

placed in different directions so as to produce
a shell which has similar structural properties
in all directions. The shell is normally
supported on edge beams of laminated timber
which are integral with the shell itself. For ease
of construction and analysis shapes which
have a regular geometry are usually adopted.
The hyperbolic paraboloid (Fig. 6.59) is particularly suitable but other forms, such as the
conoid and the elliptical paraboloid, have also
been used.

6.7.6 Lattice domes and vaults
In this form of construction triangulated
arrangements of timber elements are built up
into domed or vaulted shapes (Figs 6.60 to
6.62). The resulting structures are very
efficient due to the form-active or semi-form-

Fig. 6.62 The lamella vault is a system which allows
large interiors to be created. The degree of standardisation
which is present can make this an economical form of
construction.

active overall geometry and are therefore
suitable for large enclosures (span range
15 m to 200 m). In the lamella system each
element is twice the length of the side of a
diamond, and at each joint, one element
passes continuously through, with adjacent
intersecting elements connected to its mid-
point. Triangulation of the geometry is
effected with purlins. The particular advantage of the lattice vault or dome is the high
degree of standardisation of the components
which is possible.

Selected bibliography

Addis, W.B., *The Art of the Structural Engineer*, London, 1994

Ambrose, J., *Building Structures*, New York, 1988

Balcombe, G., *Mitchell's History of Building*, London, 1985

Baird, J. A., *Timber Specifier's Guide*, Oxford, 1990

Baird, J. A. and Ozelton, E. C., *Timber Designer's Manual*, 2nd edition, London, 1984

Benedetti, C. and Bacigulupi, V., *Legno Architettura* (with English text), Rome, 1991

Benjamin, B. S., *Structures for Architects*, 2nd edition, New York, 1984

Bill, M., *Robert Maillart*, London, 1969

Blanc, A., McEvoy, M. and Plank, R., *Architecture and Construction in Steel*, London, 1993

BS 449: Part 2: 1969: *The Use of Structural Steel in Building*, British Standards Institution, Milton Keynes

BS 6399: Part 1: 1984, *Design Loading for Buildings, Code of Practice for Dead and Imposed Loads*, British Standards Institution, Milton Keynes

Bull, J. W., *The Practical Design of Structural Elements in Timber*, 2nd edition, Aldershot, 1994

Curtin, W. G., Shaw, G., Beck, J. K. and Parkinson, G.I., *Structural Masonry Designer's Manual*, London, 1982

Curtin, W. G., Shaw, G., Beck, J. K. and Bray, W. A., *Structural Masonry Detailing*, London, 1984

Curtis, W., *Modern Architecture since 1900*, London, 1987

Desch, H. E.,*Timber* (6th edition, revised Dinwoodie, J. M.), London 1981

Dowling, P. J., Knowles, P. and Owens, G. W., *Structural Steel Design*, London, 1988

Elliott, C. D., *Technics and Architecture: The Development of Materials and Systems for Buildings*, London, 1992

Elliott, K. S., *Multi-storey Precast Concrete Frame Structures*, Oxford, 1996

Engel, H., *Structure Systems*, Stuttgart, 1967

Engel, H., *Structural Principles*, Englewood Cliffs, NJ, 1984

Friedman, D, *Historical Building Construction*, London, 1995

Gans, D. (Ed.), *Bridging the Gap: Rethinking the Relationship of Architect and Engineer*, London, 1991

Gordon, J.E., *The New Science of Strong Materials*, Harmondsworth, 1968

Gordon, J. E., *Structures*, Harmondsworth, 1978

Gorst, T., *The Buildings Around Us*, London, 1995.

Hart, F., Henn, W. and Sontag, H., *Multi-storey Buildings in Steel*, London, 1978

Hartoonian, G., *The Ontology of Construction*, Cambridge, 1994

Hayward, A. and Weare, F., *Steel Detailer's Manual*, Oxford, 1989

Hodgkinson, A. (Ed.) AJ *Handbook of Building Structure*, London, 1974

Jacobus, J., *James Stirling: Buildings and Projects*, New York, 1975

Jodidio, P., *Contemporary European Architects*, Volume 3, Cologne, 1995

Lambot, I. (Ed.), *Norman Foster-Foster Associates: Buildings and Projects: Volume 2 1971–78*, London, 1989

Lange, *Basilica at Pompeii*, Leipzig, 1885

Macdonald, A. J., *Structure and Architecture*, Oxford, 1994

Mainstone, R.J., *Developments in Structural Form*, London, 1975

Mainstone, R.J., *Hagia Sophia*, London, 1988

Mark, R, (Ed.), *Architectural Technology up to the Scientific Revolution*, Cambridge, MA, 1993

Mettem, C. J., *Structural Timber Design and Technology*, London, 1986

Noever, P. (Ed.), *The End of Architecture*, Munich, 1993

Orton, A., *The Way We Build Now*, London, 1988

Pippard, A.J.S. and Baker, J., *The Analysis of Engineering Structures* (4th edition), London, 1984.

Reynolds, C. E. and Steedman, J. C., *Reinforced Concrete Designer's Manual*, 10th edition, London, 1988

Rice, P., *An Engineer Imagines*, London, 1994

Schodek, D. L., *Structure in Sculpture*, Connecticut, 1993

Sunless, J. and Bedding, B., *Timber in Construction*, London, 1985

Thornton, C., Tomasetti, R., Tuchman, J. and Joseph, L., *Exposed Structure in Building Design*, New York, 1993

Wilkinson, C., *Supersheds*, Oxford, 1991

Yeomans, D., *The Trussed Roof: its History and Development*, Aldershot, 1992

The relationship between structural form and structural efficiency

Structural efficiency is defined here in terms of the weight of structural material which is used to provide a given load-carrying capacity. The efficiency will be taken to be high if the strength-to-weight ratio of the structure is high.

The efficiency of a structure is affected by a number of factors connected with its form and general configuration. A very significant factor is the relationship between the form of a struc-ture and the type of internal force which occurs for a given pattern of loading. Only this aspect will be discussed here.[1]

Elements in architectural structures are normally subjected to axial internal force, to bending-type internal force or to a combin-ation of these. The distinction is an important one so far as efficiency is concerned because axial internal force can be resisted more efficiently than bending-type internal force. The principal reason for this is that the distri-bution of stress which occurs within axially loaded elements is almost constant (Fig. A1.1) and this uniform level of stress allows all of the material in the element to be stressed to its limit – that is to a level which is slightly lower than the failure stress of the material. An efficient use of material therefore results because all of the material present provides full value for its weight. With bending stress, which varies in intensity in all cross-sections (Fig. A1.1) from a minimum at the neutral axis to a maximum at the extreme fibres, only the most highly stressed parts of each cross-section are used efficiently. Most of the mater-ial present is under-stressed and therefore inefficiently used.

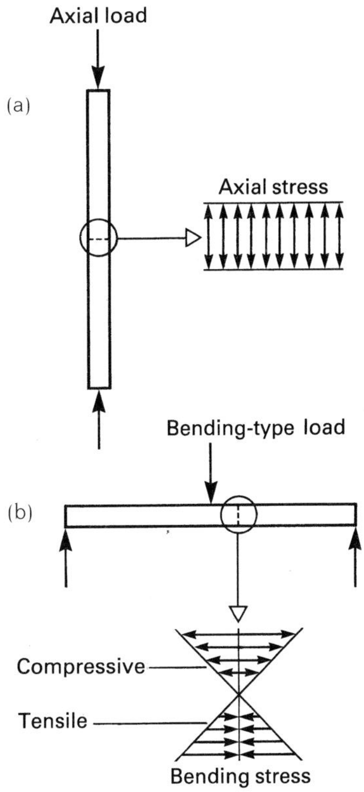

Fig. A1.1 (a) Elements which carry purely axial load are subjected to axial stress the intensity of which is constant across all cross-sectional planes.
(b) Pure bending-type load (i.e. load which is normal to the axis of the element) causes bending stress to occur on all cross-sectional planes. The magnitude of this varies within each cross-section from a maximum compressive stress at one extremity to a maximum tensile stress at the other.

The type of internal force which occurs in an element depends on the relationship between the direction of its principal axis (its longitu-dinal axis) and the direction of the load which

---

1   For a more wide-ranging discussion of structural efficiency see Macdonald, *Structure and Architecture, op. cit.*

235

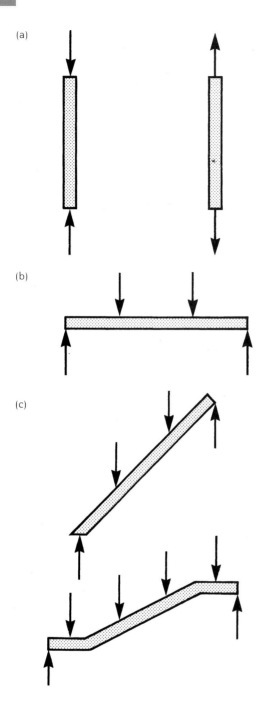

(a)

(b)

(c)

Fig. A1.2 Basic relationships between loads and structural elements.

(a) Load coincident with the principal axis; axial internal force only.

(b) Load perpendicular to the principal axis; bending-type internal force.

(c) Load inclined to the principal axis; combined axial and bending-type internal force.

is applied to it (Fig. A1.2). If an element is straight, axial internal force occurs if the load is applied parallel to the longitudinal axis of the element. Bending-type internal force occurs if it is applied at right angles to the longitudinal axis and combined axial and bending-type internal forces occur if it is applied in a direction which is inclined to the longitudinal axis. The axial-only and bending-only cases can be regarded as special cases of the more general combined case, but they are in fact the most commonly found types of loading arrangement in architectural structures.

If an element is not straight then it will almost inevitably be subjected to a combination of axial and bending internal forces when a load is applied but there are important exceptions to this as is illustrated in Fig. A1.3. Here the structural element consists of a flexible cable, supported at its ends, and from which various loads are suspended. Because the cable has no rigidity it is incapable of carrying any other type of internal force but axial tension; it is therefore forced into a shape which allows it to resist the load with an internal force which is pure axial tension. The shape traced by the longitudinal axis is unique to the load pattern and is called the *form-active* shape for that load.

As is seen in Fig. A1.3 the shape which the cable adopts is dependent on the pattern of load which is applied; the form-active shape is straight-sided when the loads are concentrated at individual points and curved if the load is distributed along it. If a cable is allowed simply to sag under its own weight, which is a distributed load acting along its entire length, it adopts a curve known as a catenary.

An interesting feature of the form-active shape for any load pattern is that if a rigid element is constructed whose longitudinal axis is the mirror image of the form-active shape taken up by the cable, then it too will be subjected only to axial internal forces when the same load is applied, despite the fact that, being rigid, it could also carry bending-type internal force. In the mirror-image form all the axial internal forces are compressive (Fig. A1.4).

The cable structure and its rigid 'mirror-image' counterpart are simple examples of a whole class of structural elements which carry axial internal forces only, due to their longitudinal axes being shaped to the form-active shapes for the loads which are applied to them. These are called *form-active elements*.

If, in a real structure, a flexible material such as steel wire or cable is used to make an element, it will automatically take up the form-active shape when the load is applied. Flexible material is in fact incapable of becoming anything other than a form-active element. If the material is rigid, however, and a form-active element is required, then it must be made to conform to the form-active shape of the load which is to be applied to it or, in the case of a compressive element, to the mirror image of the form-active shape. If not, the internal force will not be pure axial force and some bending will occur.

Figure A1.5 contains a mixture of form-active and non-form-active shapes. Two load patterns are shown: a uniformly distributed load across the whole of the element and two concentrated loads applied at equal distances across them. For each load, elements (a) carry pure bending-type internal forces; no axial force can occur in these because there is no component of either load which is parallel to the axis of the element. The elements in (b) have shapes which conform exactly to the form-active shapes of the loads. They are therefore form-active elements which carry axial internal forces only; in both cases the forces are compressive. The elements (c) do not conform to the form-active shapes of the loads and will not therefore carry pure axial internal force. Neither will they be subjected to pure bending; they will carry a combination of bending and axial internal force.

So far as the shape of their longitudinal axes are concerned, structural elements can thus be classified into three categories: form-active elements, non-form-active elements and semi-form-active elements. Form-active elements are those which conform to the form-active shape of the loads which are applied to them and they contain axial internal forces only.

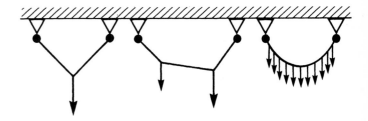

Fig. A1.3   Tensile form-active shapes. Because it has no rigidity a cable must take up a shape – the *form-active* shape – which allows it to resist the load with a purely tensile internal force. Different load arrangements produce different form-active shapes.

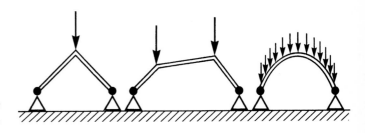

Fig. A1.4   Compressive form-active shapes.

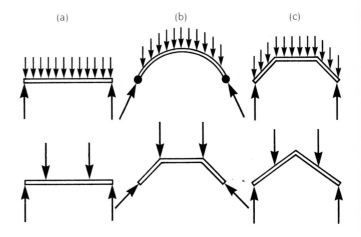

Fig. A1.5   Examples of the relationship between element shape, load pattern and element type. The latter is determined by the relationship between the shape of the element and the form-active shape for the load pattern which it carries.
(a) Non-form-active (bending stress only).
(b) Form-active (axial stress only).
(c) Semi-form-active (combined bending and axial stress).

Non-form-active elements are those whose longitudinal axis does not conform to the form-active shape of the loads and are such that no axial component of internal force occurs. These contain bending-type internal force only. Semi-form-active elements are elements whose shapes are such that they contain a combination of bending and axial internal forces. The cranked-beam shape in Fig. A1.5 is a fully-form-active element when subjected to the two concentrated loads (A1.5b) but a semi-form-

active element when subjected to the uniformly distributed load (A1.5c).

Because they are not subjected to bending stress, elements with form-active shapes are potentially the most efficient types of structure. Non-form-active elements are potentially the least efficient because bending stress is the principal type to which they are subjected. The efficiency of semi-form-active elements depends on the extent to which they differ from the form-active shape.

Approximate methods for allocating sizes to structural elements

A2.1 Introduction

The methods which are outlined here for determining the sizes of structural elements are not those of rigorous final structural design calculations. Rather they are quick methods which allow the sizes of structural elements to be determined approximately. They are preliminary planning tools which are sufficiently accurate, in most cases, to allow the feasibility of proposed structural arrangements to be assessed. They are applicable to the mainstream forms of structure.

One of the main objectives of structural design is to produce structures which are safe and serviceable at reasonable cost. This requires that the sizes of the cross-sections of structural elements be sufficiently, but only sufficiently, large to carry safely the loads which are applied to them. The principal mechanism by which this aspect of structural design is controlled is through calculations.

A2.2 Structural analysis

The principal factor which determines the size required for a structural element is the amount of load which it carries and the element-sizing part of structural calculations must normally therefore be preceded by an assessment of this. The process by which it is done is known as the analysis of the structure. Even approximate element-sizing calculations require that some form of rudimentary structural analysis be carried out.

Structural analysis can be subdivided into the three distinct processes of *load assessment*, *preliminary analysis* and *final analysis*. Load assessment involves the prediction of the maximum load which will occur on the structure in its lifetime. In the case of architectural structures there are three principal types of load: these are *dead load* – the permanent gravitational load caused by the weight of the building and its fixtures; *gravitational imposed load* – variable load caused by the weights of the occupants of the building, furniture and other moveable items; and *wind loading* – non-gravitational imposed load caused by the action of wind pressure. In the case of most structural elements the most unfavourable load grouping is the combination of dead and imposed gravitational load. The approximate element-sizing calculations which are presented here will therefore be based principally on this load combination. The figures which are specified in the current British Standard for dead and imposed gravitational loads are given in Tables A2.1 and A2.2.

The preliminary analysis of a structure is a process in which the three-dimensional object which is the structure is broken down into its constituent elements so that the forces which are imposed on the elements under the action of the maximum load condition can be determined. It involves the tracing of the path which is taken through the structure by the load from the floor and roof surfaces, to which it is applied initially, to the foundations, where it is ultimately resisted. Some structural elements, such as floor slabs, are acted on directly by the loads. Others, such as columns, receive load from the parts of the structure which they support. The end product of the preliminary analysis is a set of individual structural diagrams (one for each element) on which the forces which act on each element are marked.

Table A2.1   Weights of building materials (based on BS 648 [1 kg = 10 N (approx)])

Asphalt		*Plaster*	
Roofing 2 layers, 19 mm thick	42 kg/m²	Two coats gypsum, 13 mm thick	22 kg/m²
Damp-proofing, 19 mm thick	41 kg/m²		
Road and footpaths, 19 mm thick	44 kg/m²	*Plastics sheeting*	
		Corrugated	4.5 kg/m²
Bitumen roofing felts			
Mineral surfaced bitumen per lay	3.5 kg/m²	*Plywood*	
		per mm thick	0.7 kg/m²
Blockwork			
Solid per 25 mm thick, stone aggregate	55 kg/m²	*Reinforced concrete*	2400 kg/m³
Aerated per 25 mm thick	15 kg/m²		
		Rendering	
Board		Cement:sand (1:3) 13 mm thick	30 kg/m²
Blockboard per 25 mm thick	12.5 kg/m²		
		Screeding	
Brickwork		Cement:sand (1:3) 13 mm thick	30 kg/m²
Clay, solid per 25 mm thick, medium density	55 kg/m²		
Concrete, solid per 25 mm thick	59 kg/m²	*Slate tiles*	
		(depending upon thickness and source)	24–78 kg/m²
Cast stone	2250 kg/m³		
		Steel	
Concrete		Solid (mild)	7850 kg/m³
Natural aggregates	2400 kg/m³	Corrugated roofing sheets per mm thick	10 kg/m²
Lightweight aggregates (structural)	1760 kg/m³		
	+240 or −160	*Tarmacadam*	
		25 mm thick	60 kg/m²
Flagstones			
Concrete, 50 mm thick	120 kg/m²	*Terrazzo*	
		25 mm thick	54 kg/m²
Glass fibre			
Slab, per 25 mm thick	2.0–5.0 kg/m²	*Tiling, roof*	
		Clay	70 kg/m²
Gypsum panels and partitions			
Building panels 75 mm thick	44 kg/m²	*Timber*	
		Softwood	590 kg/m³
Lead		Hardwood	1250 kg/m³
Sheet, 2.5 mm thick	30 kg/m²		
		Water	1000 kg/m³
Linoleum			
3 mm thick	6 kg/m²	*Woodwool*	
		Slabs, 25 mm thick	15 kg/m²

The final part of the analysis of a structure consists of the evaluation of the internal forces in the individual elements – the bending moments, shear forces and axial forces which act on their cross-sections. It is the magnitudes of these which determine the shapes and sizes which are required for the cross-sections. The analytical technique by which they are calculated is the device of the 'imaginary cut' through the structure, which is used in conjunction with the concept of equilibrium and the 'free-body-diagram' to evaluate the internal forces (see Macdonald, *Structure and Architecture*, Chapter 2 and Appendix 1, for an explanation of these terms).

In the context of approximate calculations for preliminary element sizing a much abbreviated version of structural analysis can be carried out. Often, the requirement is to determine approximately the sizes of the critical elements only (e.g. the most heavily loaded element in a timber roof truss or the most

Table A2.2   Minimum imposed floor loads (based on BS 6399 Part 1 1996)

Type of activity/occupancy for part of the building or structure	Examples of specific use		Uniformly distributed load kN/m²	Concentrated load kN
A Domestic and residential activities (Also see category C)	All usages within self-contained dwelling units Communal areas (including kitchens) in blocks of flats with limited use (See note 1) (For communal areas in other blocks of flats, see C3 and below)		1.5	1.4
	Bedrooms and dormitories except those in hotels and motels		1.5	1.8
	Bedrooms in hotels and motels Hospital wards Toilet areas		2.0	1.8
	Billiard rooms		2.0	2.7
	Communal kitchens except in flats covered by note 1		3.0	4.5
	Balconies	Single dwelling units and communal areas in blocks of flats with limited use (See note 1)	1.5	1.4
		Guest houses, residential clubs and communal areas in blocks of flats except as covered by note 1	Same as rooms to which they give access but with a minimum of 3.0	1.5/m run concentrated at the outer edge
		Hotels and motels	Same as rooms to which they give access but with a minimum of 4.0	1.5/m run concentrated at the outer edge
B Offices and work areas not covered elsewhere	Operating theatres, X-ray rooms, utility rooms		2.0	4.5
	Work rooms (light industrial) without storage		2.5	1.8
	Office for general use		2.5	2.7
	Banking halls		3.0	2.7
	Kitchens, laundries, laboratories		3.0	4.5
	Rooms with mainframe computers or similar equipment		3.5	4.5
	Machinery halls, circulation spaces therein		4.0	4.5
	Projection rooms		5.0	To be determined for specific use
	Factories, workshops and similar buildings (general industrial)		5.0	4.5
	Foundries		20.0	To be determined for specific use
	Catwalks		—	1.0 at 1 m centres
	Balconies		Same as rooms to which they give access but with a minimum of 4.0	1.5/m run concentrated at the outer edge
	Fly galleries		4.5 kN/m run distributed uniformly over width	—
	Ladders		—	1.5 rung load

Table A2.2 Selected gravitational imposed loads (*continued*)

Type of activity/occupancy for part of the building or structure	Examples of specific use		Uniformly distributed load kN/m²	Concentrated load kN
C Areas where people may congregate	Public, institutional and communal dining rooms and lounges, cafes and restaurants (See note 2)		2.0	2.7
C1 Areas with tables	Reading rooms with no book storage		2.5	4.5
	Classrooms		3.0	2.7
C2 Areas with fixed seats	Assembly areas with fixed seating (See note 3)		4.0	3.6
	Places of worship		3.0	2.7
C3 Areas without obstacles for moving people	Corridors, hallways, aisles, stairs, landings etc. in institutional type buildings (not subject to crowds or wheeled vehicles), hostels, guest houses, residential clubs, and communal areas in blocks of flats not covered by note 1. (For communal areas in blocks of flats covered by note 1, see A)	Corridors, hallways, aisles etc. (foot traffic only)	3.0	4.5
		Stairs and landings (foot traffic only)	3.0	4.0
	Corridors, hallways, aisles, stairs, landing, etc. in all other buildings including hotels and motels and institutional buildings	Corridors, hallways, aisles etc. (foot traffic only)	4.0	4.5
		Corridors, hallways, aisles etc., subject to wheeled vehicles, trolleys etc.	5.0	4.5
		Stairs and landings (foot traffic only)	4.0	4.0
	Industrial walkways (light duty)		3.0	4.5
	Industrial walkways (geneal duty)		5.0	4.5
	Industrial walkways (heavy duty)		7.5	4.5
	Museum floors and art galleries for exhibition purposes		4.0	4.5
	Balconies (except as specified in A)		Same as rooms to which they give access but with a minimum of 4.0	1.5/m run concentrated at the outer edge
	Fly galleries		4.5 kN/m run distributed uniformly over width	—

Table A2.2 Selected gravitational imposed loads (*continued*)

Type of activity/occupancy for part of the building or structure	Examples of specific use	Uniformly distributed load kN/m²	Concentrated load kN
C4 Areas with possible physical activities (See clause **9**)	Dance halls and studios, gymnasia, stages	5.0	3.6
	Drill halls and drill rooms	5.0	9.0
C5 Areas susceptible to overcrowding (See clause **9**)	Assembly areas without fixed seating, concert halls, bars, places of worship and grandstands	5.0	3.6
	Stages in public assembly areas	7.5	4.5
D Shopping areas	Shop floors for the sale and display of merchandise	4.0	3.6
E Warehousing and storage areas. Areas subject to accumulation of goods. Areas for equipment and plant.	General areas for static equipment not specified elsewhere (institutional and public buildings)	2.0	1.8
	Reading rooms with book storage, e.g. libraries	4.0	4.5
	General storage other than those specified	2.4 for each metre of storage height	7.0
	File rooms, filing and storage space (offices)	5.0	4.5
	Stack rooms (books)	2.4 for each metre of storage height but with a minimum of 6.5	7.0
	Paper storage for printing plants and stationery stores	4.0 for each metre of storage height	9.0
	Dense mobile stacking (books) on mobile trolleys, in public and institutional buildings	4.8 for each metre of storage height but with a minimum of 9.6	7.0
	Dense mobile stacking (books) on mobile trucks, in warehouses	4.8 for each metre of storage height but with a minimum of 15.0	7.0
	Cold storage	5.0 for each metre of storage height but with a minimum of 15.0	9.0
	Plant rooms, boiler rooms, fan rooms, etc., including weight of machinery	7.5	4.5
	Ladders	—	1.5 rung load
F	Parking for cars, light vans, etc. not exceeding 2500 kg gross mass, including garages, driveways and ramps	2.5	9.0
G	Vehicles exceeding 2500 kg. Driveways, ramps, repair workshops, footpaths with vehicle access, and car parking	To be determined for specific use	

NOTE 1. Communal areas in blocks of flats with limited use refers to blocks of flats not more than three storeys in height and with not more than four self-contained dwelling units per floor accessible from one staircase.

NOTE 2. Where these same areas may be subjected to loads due to physical activities or overcrowding, e.g. a hotel dining room used as a dance floor, imposed loads should be based on occupancy C4 or C5 as appropriate. Reference should also be made to clause **9** (BS 6399 Part 1: 1996)

NOTE 3. Fixed seating is seating where its removal and the use of the space for other purposes is improbable.

Table A2.3 Bending moment and deflection formulae for beams (after AJ Handbook of Building Structure)

The following tabulates formulae and values of moments (M), reactions (R), shear force (S) and deflection (δ) in beams for a number of common loading and support conditions. It covers cantilevers, free support beams, fixed-end beams, and propped cantilevers

I Cantilevers

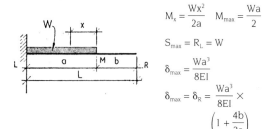

$$M_x = \frac{Wx^2}{2a} \quad M_{max} = \frac{Wa}{2}$$

$$S_{max} = R_L = W$$

$$\delta_{max} = \frac{Wa^3}{8EI}$$

$$\delta_{max} = \delta_R = \frac{Wa^3}{8EI} \times$$

$$\left(1 + \frac{4b}{3a}\right)$$

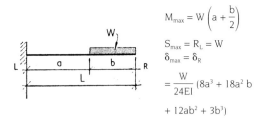

$$M_{max} = W\left(a + \frac{b}{2}\right)$$

$$S_{max} = R_L = W$$
$$\delta_{max} = \delta_R$$

$$= \frac{W}{24EI}(8a^3 + 18a^2 b$$

$$+ 12ab^2 + 3b^3)$$

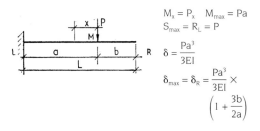

$$M_x = P_x \quad M_{max} = Pa$$
$$S_{max} = R_L = P$$

$$\delta = \frac{Pa^3}{3EI}$$

$$\delta_{max} = \delta_R = \frac{Pa^3}{3EI} \times$$

$$\left(1 + \frac{3b}{2a}\right)$$

2 Free support beams

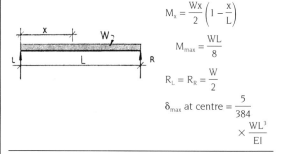

$$M_x = \frac{Wx}{2}\left(1 - \frac{x}{L}\right)$$

$$M_{max} = \frac{WL}{8}$$

$$R_L = R_R = \frac{W}{2}$$

$$\delta_{max} \text{ at centre} = \frac{5}{384}$$

$$\times \frac{WL^3}{EI}$$

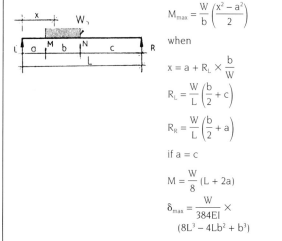

$$M_{max} = \frac{W}{b}\left(\frac{x^2 - a^2}{2}\right)$$

when

$$x = a + R_L \times \frac{b}{W}$$

$$R_L = \frac{W}{L}\left(\frac{b}{2} + c\right)$$

$$R_R = \frac{W}{L}\left(\frac{b}{2} + a\right)$$

if a = c

$$M = \frac{W}{8}(L + 2a)$$

$$\delta_{max} = \frac{W}{384EI} \times$$

$$(8L^3 - 4Lb^2 + b^3)$$

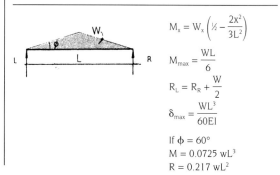

$$M_x = W_x\left(\frac{1}{2} - \frac{2x^2}{3L^2}\right)$$

$$M_{max} = \frac{WL}{6}$$

$$R_L = R_R + \frac{W}{2}$$

$$\delta_{max} = \frac{WL^3}{60EI}$$

If φ = 60°
M = 0.0725 wL³
R = 0.217 wL²

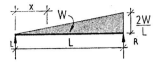

$$M_x = \frac{W_x}{3}\left(1 - \frac{x^2}{L^2}\right)$$

$$M_{max} = 0.128WL$$
$$X_1 = 0.5774L$$

$$R_L = \frac{W}{3}$$

$$R_R = \frac{2W}{3}$$

$$\delta_{max} = \delta x_2 = \frac{0.01304WL^3}{EI}$$

$$X_2 = 0.5193L$$

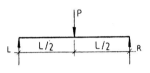

$$M_{max} = \frac{PL}{4}$$

$$R_L = R_R = \frac{P}{2}$$

$$\delta_{max} = \frac{PL^3}{48EI}$$

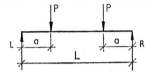

$$M_{max} = Pa$$
$$R_L = R_R = P$$

$$\delta_{max} = \frac{PL3}{6EI}\left[\frac{3a}{4L} - \left(\frac{a}{L}\right)^3\right]$$

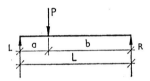

$$M_{max} = P\frac{ab}{L} = M_p$$

$$R_L = \frac{Pb}{L} \quad R_R = \frac{Pa}{L}$$

δ_{max} always occurs within 0.0774L of the centre of the beam. When b > a

$$\delta \text{ centre} = \frac{PL^3}{48EI} \times$$
$$\left[3\frac{a}{L} - 4\left(\frac{a}{L}\right)^3\right]$$

This value is always within 2.5% of the maximum value.

$$\delta_p = \frac{PL^3}{3EI}\left(\frac{a}{L}\right)^2\left(1 - \frac{a}{L}\right)^2$$

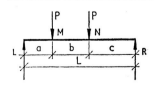

$$M_M = \frac{Pa(b + 2c)}{2L}$$

$$M_N = \frac{Pc(b + 2a)}{2L}$$

$$R_L = \frac{P(b + 2c)}{L}$$

$$R_R = \frac{P(b + 2a)}{L}$$

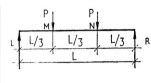

$$M_{max} = \frac{PL}{3}$$

$$R_L = R_R = P$$

$$\delta_{max} = \frac{23\,PL^3}{648EI}$$

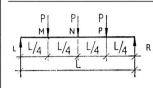

$$M_{max} = M_N = \frac{PL}{2}$$

$$M_M = M_P = \frac{3PL}{8}$$

$$R_L = R_R = \frac{3P}{2}$$

$$\delta_{max} = \frac{19\,PL^a}{384EI}$$

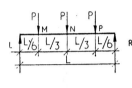

$$M_{max} = \frac{5PL}{12}$$

$$M_M = M_P = \frac{PL}{4}$$

$$R_L = R_R = \frac{3P}{2}$$

$$\delta_{max} = \frac{53\,PL^3}{12964EI}$$

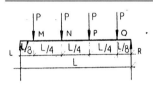

$$M_{max} = P_N = P_P = \frac{PL}{2}$$

$$M_M = M_Q = \frac{PL}{4}$$

$$R_L = R_R = 2P$$

$$\delta_{max} = \frac{41\,PL^3}{768EI}$$

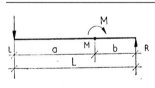

$$M_{ML} = M\frac{a}{L} \quad M_{MR} = M\frac{b}{L}$$

$$R_A = R_B = \frac{M}{L}$$

when a > b

$$\delta_m = -\frac{Mab}{3EI}\left(\frac{a}{L} - \frac{b}{L}\right)$$

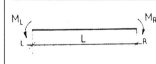

$$R_L = -R_R = \frac{M_L - M_R}{L}$$

when $M_L = M_R$,

$$\delta_{max} = -\frac{ML^2}{8EI}$$

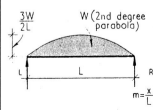

W (2nd degree parabola)

$$M_x = \frac{WL}{2}(m^4 - 2m^3 + m)$$

$$M_{max} = \frac{5WL}{32}$$

$$R_L = R_R = \frac{W}{2}$$

$$m = \frac{x}{L} \quad \delta_{max} = \frac{6.1\,WL^3}{384\,EI}$$

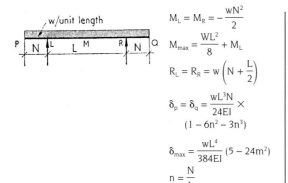

$$M_L = M_R = -\frac{wN^2}{2}$$

$$M_{max} = \frac{WL^2}{8} + M_L$$

$$R_L = R_R = w\left(N + \frac{L}{2}\right)$$

$$\delta_p = \delta_q = \frac{wL^3N}{24EI} \times (1 - 6n^2 - 3n^3)$$

$$\delta_{max} = \frac{wL^4}{384EI}(5 - 24m^2)$$

$$n = \frac{N}{L}$$

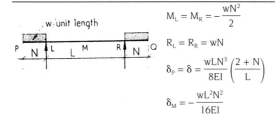

$$M_L = M_R = -\frac{wN^2}{2}$$

$$R_L = R_R = wN$$

$$\delta_p = \delta = \frac{wLN^3}{8EI}\left(\frac{2 + N}{L}\right)$$

$$\delta_M = -\frac{wL^2N^2}{16EI}$$

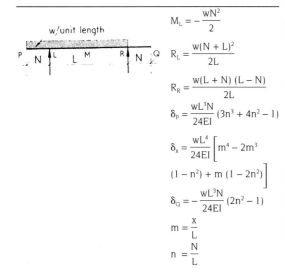

$$M_L = -\frac{wN^2}{2}$$

$$R_L = \frac{w(N + L)^2}{2L}$$

$$R_R = \frac{w(L + N)(L - N)}{2L}$$

$$\delta_p = \frac{wL^3N}{24EI}(3n^3 + 4n^2 - 1)$$

$$\delta_x = \frac{wL^4}{24EI}\left[m^4 - 2m^3(1 - n^2) + m(1 - 2n^2)\right]$$

$$\delta_Q = -\frac{wL^3N}{24EI}(2n^2 - 1)$$

$$m = \frac{x}{L}$$

$$n = \frac{N}{L}$$

3 Fixed-end beams

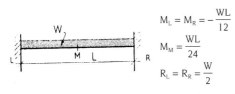

$$M_L = M_R = -\frac{WL}{12}$$

$$M_M = \frac{WL}{24}$$

$$R_L = R_R = \frac{W}{2}$$

points of contraflexure
0.21L from each end

$$\delta_{max} = \frac{WL^3}{384EI}$$

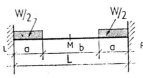

$$M_L = M_R = -\frac{Wa}{12L} \times (3L - 2a)$$

$$M_M = \frac{Wa}{4} + M_L$$

$$R_L = R_R = \frac{W}{2}$$

$$\delta_{max} = \frac{Wa^2}{48EI}(L - a)$$

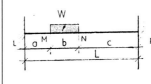

$$M_L = -\frac{W}{12L^2b} \times \left[e^3(4 - 3e) - c^3(4L - 3c)\right]$$

$$M_R = -\frac{W}{12L^2b} \times \left[d^3(4L - 3d) - a^3(4L - 3a)\right]$$

a + b = d
b + c = e
when r = reaction if the
beam were simply
supported,

$$R_L = r_L + \frac{M_L - M_R}{L}$$

$$R_R = r_R + \frac{M_R - M_L}{L}$$

when a = c,

$$\delta_{max} = \frac{W}{384EI} \times (L^3 + 2L^2a + 4La^2 - 8a^3)$$

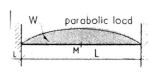

$$M_L = M_R = -\frac{WL}{10}$$

$$M_M = \frac{5WL}{32} - \frac{WL}{10} = \frac{9WL}{160}$$

$$R_L = R_B = \frac{W}{2}$$

$$\delta_{max} = \frac{1.3\,WL^3}{384\,EI}$$

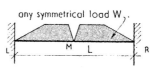

$$M_L = M_R = -\frac{A_S}{L}$$

where A_S is the area of
the free bending
moment diagram

$$R_L = R_R = \frac{W}{2}$$

$$\delta_{max} = \frac{A_S x - A_1 x_1}{2EI}$$

HALF
BENDING
MOMENT
DIAGRAM

centre of area of
half fixed end BMD

centre of area of
half free end BMD

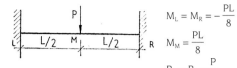

$$M_L = M_R = -\frac{PL}{8}$$

$$M_M = \frac{PL}{8}$$

$$R_L = R_R = \frac{P}{2}$$

$$\delta_{max} = \frac{PL_3}{192EI}$$

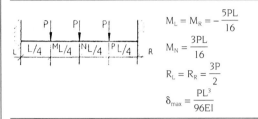

$$M_L = M_R = -\frac{19PL}{72}$$

$$M_N = \frac{11PL}{72}$$

$$R_L = R_R = \frac{3P}{2}$$

$$\delta_{max} = \frac{41PL^3}{5184EI}$$

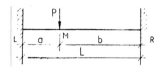

$$M_L = -\frac{Pab^2}{L^2}$$

$$M_R = -\frac{Pba^2}{L^2}$$

$$M_M = -\frac{2Pa^2b^2}{L^3}$$

$$R_L = P\frac{b^2}{L^2}\left(1 + 2\frac{a}{L}\right)$$

$$R_R = P\frac{a^2}{L^2}\left(1 + 2\frac{b}{L}\right)$$

$$\delta_M = \frac{Pa^3b_3}{3EIL_3}$$

$$\delta_{max} = \frac{2Pa^2b^3}{3EIL\,(3L - 2a)^2}$$

$$\text{at } x = \frac{L^2}{3L - 2a}$$

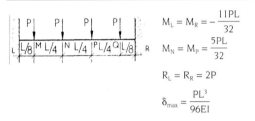

$$M_L = M_R = -\frac{5PL}{16}$$

$$M_N = \frac{3PL}{16}$$

$$R_L = R_R = \frac{3P}{2}$$

$$\delta_{max} = \frac{PL^3}{96EI}$$

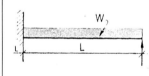

$$M_L = M_R = -\frac{11PL}{32}$$

$$M_N = M_P = \frac{5PL}{32}$$

$$R_L = R_R = 2P$$

$$\delta_{max} = \frac{PL^3}{96EI}$$

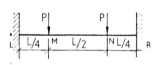

$$M_L = M_R = -\frac{3PL}{16}$$

$$M_M = M_N = \frac{PL}{16}$$

$$R_L = R_R = P$$

$$\beta max = \frac{PL^3}{192EI}$$

4 Propped cantilevers

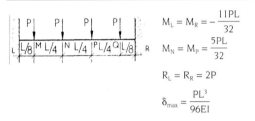

$$M_L = -\frac{WL}{8}$$

$$M_{max} = \frac{9WL}{128} \text{ at } x' = \tfrac{5}{8}$$

$$M = 0 \text{ at } x' = \tfrac{1}{4}$$

$$R_L = \tfrac{5}{8}\,W$$

$$R_R = \tfrac{3}{8}\,W$$

$$\text{if } m = 1 - x'$$

$$\delta = \frac{WL^3}{48EI} \times$$

$$(m - 3m^3 + 2m^4)$$

$$\delta_{max} = \frac{WL^3}{185EI}$$

$$\text{at } x' = 0.5785$$

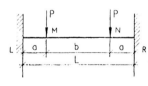

$$M_L = M_R = -\frac{Pa(L - a)}{L}$$

$$M_M = M_N = \frac{Pa^2}{L}$$

$$R_L = R_R = P$$

$$\delta_{max} = \frac{PL^3}{6EI}\left[\frac{3a^2}{4L^2} - \frac{a^3}{L^3}\right]$$

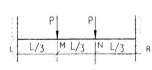

$$M_L = M_R = -\frac{2PL}{9}$$

$$M_M = M_N = \frac{PL}{9}$$

$$R_L = R_R = P$$

$$\delta_{max} = \frac{5PL^3}{648EI}$$

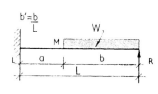

$$M_L = -\frac{Wb}{8}(2 - b^{12})$$

$$M_M = \frac{Wb}{8}(6b' - b^{13} - 4)$$

$$M = 0 \text{ where } x' = \frac{2 - b^{12}}{6 - b^{12}}$$

$$R_L = \frac{Wb'}{8}(6 - b^{12})$$

$$R_R = \frac{W}{8}(b^{13} - 6b' + 8)$$

if $x < a$,

$$\delta = \frac{WbL^2}{48EI} \times$$

$$\left[(b^{12} - 6)x^{13} - (3b^{12} - 6)x^{12}\right]$$

if $x > a$

$$\delta = \frac{WL^4}{48bEI} \times$$

$$\left[2p^4 - p^3 b'(b^{13} - 6b' + 8) + pb^{12}(3b^{12} - 8b' + 6)\right]$$

where $p = 1 - x'$

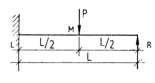

$$M_L = -\frac{W}{8L^2B}(d^2 - c^2) \times$$

$$(2L^2 - c^2 - d^2)$$

where $d = b + c$

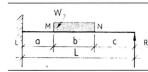

$$M_L = -\frac{3}{16}PL$$

$$M_M = \frac{5}{32}PL$$

$$R_L = \frac{11}{16}P$$

$$R_R = \frac{5}{16}P$$

$$\delta_m = \frac{7PL^3}{768EI}$$

$$\delta_{max} = 0.00932\frac{PL^3}{EI}$$

at $x' = 0.553$

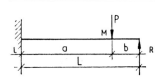

$$M_L = -\frac{Pb}{2}(1 - b^{12})$$

(maximum 0.193PL if b' = 0.577)

$$M_M = \frac{Pb}{2}(2 - 3b' + b^{13})$$

(maximum 0.174PL if b' = 0.366)

$$R_R = \frac{1}{2} Pa^{12}(b' + 2)$$

$$\delta_m = \frac{Pa^3 b^2}{12EI \, L^3} - (4L - a)$$

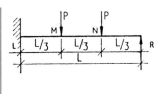

$$M_L = -\frac{1}{3}PL$$

$$M_M = \frac{1}{9}PL$$

$$M_N = \frac{2}{9}PL$$

$$R_L = \frac{4}{3}P$$

$$R_R = \frac{2}{3}P$$

$$\delta_{max} = 0.0152\frac{PL^3}{EI}$$

at $x' = 0.577$

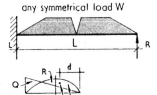

$$M_L = -\frac{19PL}{48}$$

$$M_N = \frac{21PL}{96}$$

$$M_P = \frac{53PL}{288}$$

$$R_L = \frac{91P}{48}$$

$$R_R = \frac{53P}{48}$$

$$\delta_{max} = 0.0169\frac{PL^3}{EI}$$

at $x' = 0.577$

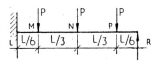

any symmetrical load W

centre of gravity of S

If A_S = area of free bending moment diagram

$$M_L = \frac{3A_S}{2L}$$

$$R_L = \frac{W}{2} + \frac{M_L}{L}$$

$$R_R = \frac{W}{2} - \frac{M_L}{L}$$

δ_{max} at X where area Q = area R

$$\delta_{max} = \frac{\text{area S} \times X \times d}{EI}$$

heavily loaded beam in a floor) in order that the feasibility of a proposal can be tested or the depth of the structural zone required for a floor estimated. In the case of elements loaded in bending, Table A2.3 can be used to estimate the maximum bending moment involved.

The loading on critical elements can be estimated from a judgement of the total area of roof or floor which they support. These areas are then simply multiplied by the intensity of gravitational load to give the load on the structural element concerned.

A2.3 Element-sizing calculations

A2.3.1 Introduction

Structural elements must be of adequate strength and must not undergo excessive deflection under the action of load. These are distinct requirements and either one of them may be the critical factor which determines the size of cross-section which must be adopted. The calculations which are outlined here are almost exclusively strength calculations. In the majority of cases these are sufficient for the purpose of determining approximately the size of cross-section required for a structural element. The basic principles and elementary forms of element-sizing calculations are outlined in this section. The modified versions of these which are used for the principal structural materials, and which allow for the peculiarities of different materials, are discussed in Sections A2.4 to A2.7

Element-sizing calculations must have a factor of safety incorporated into them to allow for the uncertainties which are inevitably present in the design and construction processes. The agencies which give rise to these inaccuracies include the imprecision with which loads and material strengths can be known, the inaccuracies which are present in the calculations themselves and the discrepancies which occur between the sizes and strengths which are specified for structural elements and those which are actually realised in the finished structure.

Basic versions of element-sizing calculations are normally based on either the *permissible stress* or the *load factor* method of design. In the permissible stress method the failure stress of the material is divided by the factor of safety to give a permissible stress and the calculations are used to determine the sizes of cross-sections required to ensure that the actual stress in the structure, when the peak load is applied, is never greater than the permissible stress.

In the load factor method the peak load value is multiplied by the factor of safety to give a factored design load which is then used, in conjunction with the actual failure stress of the material (normally taken to be the yield stress), to determine safe values for the sizes of cross-sections (i.e. values which ensure that the structure will have a margin of strength over that which is required to resist the peak values of the applied loads).

In some present-day final design calculations it is normal to split the factor of safety into two or more 'partial factors of safety' which are applied separately to load and material strength values to take account of the varying degrees of precision with which these can be known. The relative advantages and disadvantages of the different approaches to the incorporation of the factor of safety into structural calculations will not be discussed here. All of the approximate sizing calculations presented here are based on the permissible stress method, which is the simplest to apply in practice.

A2.3.2 Elements subjected to axial tension

Elements which are subjected to axial tension are normally constructed either of steel or of timber. The axial tensile stress in the element is normally considered to be uniformly distributed across the cross-section and is calculated from the equation,

$$f_{at} = P/A \qquad (A2.1)$$

where: f_{at} = axial stress
P = applied axial force
A = area of cross-section

249

If the size of cross-section does not vary along the length of an element the magnitude of the stress is the same at all locations.

The size of cross-section which is required for a particular tensile element is calculated from the following variation of the above equation:

$$A_{req} = P/f_{atp} \qquad (A2.2)$$

where: A_{req} = area of cross-section required
  P = applied axial load
  f_{atp} = maximum permissible axial tensile stress

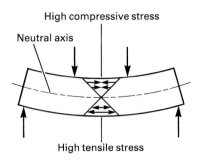

Fig. A2.1 Distribution of bending stress. Where bending-type load is present the resulting bending stress on each cross-section varies from maximum tension on one side to maximum compression on the other.

The area of cross-section which is required in practice is normally slightly larger than that given by (A2.2) due to complications with the end connection.[1] In particular, the need to allow for the effect of stress concentrations around fastening elements, such as bolts and screws, and for any eccentricity which may be present in the connection (such as might occur where a steel angle element is connected through one leg only) normally require that a slightly larger cross-section be adopted than is given by equation (A2.2). For the purpose of approximate element sizing the adoption of the very crude device of simply increasing the size of the cross-section calculated from equation (A2.2) by 15% will normally give a satisfactory result.

A2.3.3 Elements subjected to bending

A2.3.3.1 Calculation of bending stress
Bending stress occurs in an element if the external loads cause bending moment to act on its cross-sections. The magnitude of the bending stress varies within each cross-section from peak values in tension and compression in the extreme fibres on opposite sides of the cross-section, to a minimum stress in the

centre (at the centroid) where the stress changes from compression to tension (Fig. A2.1). It will normally also vary between cross-sections due to variation in the bending moment along the length of the element.

The magnitude of bending stress at any point in an element depends on the following four factors: the bending moment at the cross-section in which the point is situated; the size of the cross-section; the shape of the cross-section; and the location of the point within the cross-section. The relationship between these parameters is,

$$f_{by} = My/I \qquad (A2.3)$$

where f_{by} = bending stress at a distance y from the neutral axis of the cross-section (the axis through the centroid)
  M = bending moment at the cross-section
  I = the second moment of area of the cross-section about the axis through its centroid; this depends on both the size and the shape of the cross-section.

This relationship allows the bending stress at any location in any element cross-section to be calculated from the bending moment at that cross-section. It is equivalent to the axial stress formula $f_a = P/A$.

---

1 The making of satisfactory tensile connections is one of the classic problems of structural engineering. See Gordon, J. E., *Structures*, Harmondsworth, 1978, Section 2 for a discussion of this.

Equation (A2.3) is called the elastic bending formula. It is only valid if the peak stress is within in the elastic range. It is one of the most important relationships in the theory of structures and it is used, in a variety of forms, in the design calculations of all structural elements which are subjected to bending-type loads.

For the purpose of calculating the maximum bending stress, which occurs at the extreme fibres of the cross-section, equation (A2.3) is frequently written in the form,

$$f_{by\ max} = M/Z \tag{A2.4}$$

where: $Z = I/y_{max}$

Z is called the modulus of the cross-section. (It is often referred to as the 'section modulus'; sometimes the term 'elastic modulus' is used and this is unfortunate because it leads to confusion with the term modulus of elasticity.) If the cross-section of an element is not symmetrical about the axis through its centroid the maximum stresses in tension and compression are different. Where this occurs two section moduli are quoted for the cross-section, one for each value of y_{max}.

A2.3.3.2 Calculation of shear stress
Shear stress acts on the cross-sectional planes of bending elements due to the presence of shear force. The distribution of shear stress within a cross-section is not uniform (the pattern of distribution depends on the shape of the cross-section) but normally only the average value of shear stress is calculated.

average shear stress = shear force/area of cross-section which resists shear

$$v = V/A_v \tag{A2.5}$$

In the case of a rectangular cross-section the area which resists shear is the total area of the cross-section. For I- and box-sections the area of the web only is used.

A2.3.3.3 Approximate sizing of bending-type elements
Bending-type elements are subjected to both bending and shear stress and the size of cross-section which is adopted must be such that neither is excessively large. Normally, the bending strength criterion is used for initial selection of the cross-section size and the chosen section is then checked to ensure that it will be satisfactory in respect of shear.

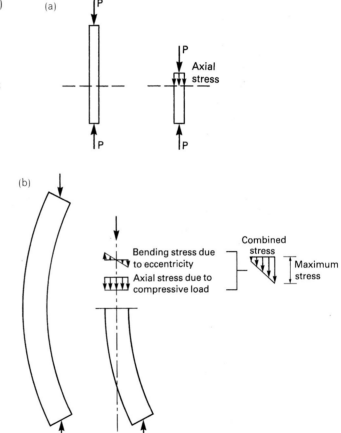

Fig. A2.2 Stresses in compression elements.
(a) If the element is straight and perfectly aligned with the load the stress in each cross-section is axial.
(b) If the element has a slight curvature the eccentricity which is present gives rise to bending stress and the total stress is a combination of this and axial stress.

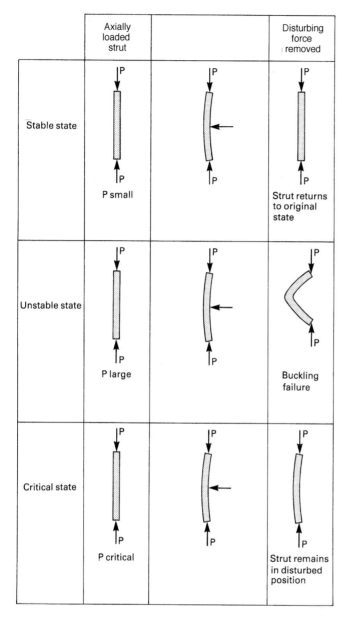

	Axially loaded strut		Disturbing force removed
Stable state	P small		Strut returns to original state
Unstable state	P large		Buckling failure
Critical state	P critical		Strut remains in disturbed position

Fig. A2.3 Stability of a 'perfect' strut (i.e. a compression element which is straight initially and perfectly aligned with the load). The condition with respect to stability depends on the level of applied axial load. If this is small, as in the first diagram, the strut is stable and will return to the original condition following a disturbance. If the axial load is high the strut is unstable and will buckle if disturbed. The third diagram depicts the situation when the 'critical' level of load, which occurs at the transition between these two load ranges, is applied. When the critical load is applied the strut remains in the displaced position following a disturbance.

Initial selection of section size can be based on the following version of the elastic bending formula:

$$Z_{req} = M/f_{bp} \qquad (A2.6)$$

where: Z_{req} = required modulus of section
M = maximum applied bending moment
f_{bp} = permissible bending stress

A2.3.4 Sizing of elements subjected to axial compression

Compression elements are the most problematic to size due to the need to allow for the phenomenon of buckling. Compressive forces are inherently unstable because if any eccentricity is present in a compressive system the action of the forces causes the amount of eccentricity to increase. For example, in Fig. A2.2 the internal forces in a simple compressive element are exposed by use of the device of the imaginary cut. It can be seen from (a) that the internal force is one of pure axial compression if the element is perfectly straight; an axial stress is produced which is evenly distributed across the cross-section and the system is in a state of equilibrium. It is, however, potentially unstable.

If the element is given slight curvature, as in (b), which might occur due to the presence of a small lateral disturbance, a couple is produced by the misalignment of the internal axial forces. This causes a bending moment to act on each cross-section. The bending strain which the curvature generates produces bending stress which resists the bending moment and which tends to restore the element to its original straight condition. Unlike in a beam, however, the bending moment and the bending stress which occur when curvature is introduced into a compressive element are not directly related. The bending moment is dependent solely on the magnitude of the applied compressive force and the amount of eccentricity which is present. The bending stress is determined by the bending strain (dependent on the amount of curvature which has developed) and on the properties of the element itself (specifically, on the geometric properties of its cross-section and on the modulus of elasticity of the material). For a particular amount of curvature (i.e. of bending strain), the size of the restoring couple generated by the bending stress is always the same, but the size of the disturbing couple caused by the eccentricity depends on the magnitude of the compressive load. Compressive elements are therefore potentially unstable internally depending on the relationship between these couples. The bending type failure which results from this type of instability is known as buckling.

The critical factor in determining the susceptibility of a particular element to buckling, is the magnitude of the applied compressive load (Fig. A2.3). If this is small the disturbing couple will also be small, even if the eccentricity is large, and the restoring couple will increase at a faster rate than the disturbing couple if some sideways-acting external agency destroys the original straight alignment of the element. The system is stable because the restoring couple will always be able to return the element to its original straight condition when the external agency is removed. If the compressive load is large, however, this will produce a disturbing couple which is greater than the restoring couple at all levels of curvature and the element will be unstable because any external agency which introduces a small amount of curvature will precipitate a progressive increase in curvature until buckling failure occurs. For a particular compressive element the magnitude of the compressive load at which instability develops is known as the critical load [P_{cr}].

The analysis of buckling is one of the classic problems of structural design and many mathematicians and engineers have investigated methods for predicting the critical loads of structural components. Perhaps the best known of these is due to the eighteenth-century Swiss mathematician Leonhard Euler and, although the formula which Euler derived for the calculation of critical loads is not suitable for most practical designs, it is nevertheless described here because the study of it provides a good introduction to the factors on which the stability of compressive elements depends.

Euler's analysis of buckling
Euler's analysis, which is described in detail in Pippard and Baker, *The Analysis of Engineering Structures* (4th Edition, Edward Arnold, London, 1984), is a theoretical investigation of a perfect strut, that is of a compressive element which is perfectly straight initially and in which no eccentricity is present, either in the element itself or in the application of the load. It yields the following formula for the critical load of a perfect strut with hinged end connections,

$$P_{cr} = \pi^2 EI/L^2 \qquad (A2.7)$$

where P_{cr} = the Euler critical load
E = the modulus of elasticity of the material
I = the second moment of area of the cross-section of the strut
L = the length of the strut

The Euler critical load for an ideal strut is equivalent to the buckling strength of a real strut. In the case of the ideal strut, a curvature must be deliberately introduced to cause buckling failure and the compressive load concerned must be greater than the critical load. Eccentricity is always present in a real strut, however, and so if a compressive load greater than the critical load is applied to it the strut will automatically fail by buckling.

It will be seen that Euler identified the slenderness of a compressive element as the most significant factor which determines the critical load, with slenderness being defined, in the most basic version of the formula as L^2/I. The more slender the element, i.e. the higher the slenderness ratio, the smaller is the critical load.

It is significant that the quantity by which the 'thickness' of the element is judged in the determination of its slenderness is the second moment of area of the cross-section (I). This is, of course, a measure of the bending performance of the cross-section and its use in this context of compressive stability is not surprising because buckling is a bending phenomenon.

The effect of second moment of area on the buckling characteristics of an element can be seen by considering the behaviour of elements with different shapes of cross-section. If the conditions of lateral restraint are the same in all directions and an increasing amount of axial load is applied to an element, it will always buckle about the axis through the centroid of its cross-section about which the bending strength is least. This is the axis which gives the smallest second moment of area. In Fig. A2.4, for example, the strut with the rectangular cross-section will fail by compressive instability at a lower load than that with the square cross-section, even though their total cross-sectional areas are the same, because the second moment of area of the rectangular cross-section, about one axis through its centroid, is very small. The fact that the second moments of area of its cross-section about other axes are larger than those of the strut with the square cross-section does not affect its compressive strength since it buckles about its weakest axis.

Because the quantity second moment of area must always be calculated with respect to a particular axis through the cross-section of an element, the critical load which is calculated by the Euler formula applies to buckling in a particular plane. This is the plane which is normal to the axis which is used to calculate I (Fig. A2.4). For a given element cross-section, a number of different critical loads can often be calculated from the Euler formula depending on the axis which is chosen for the I value. Each relates to a particular plane of buckling. If the conditions of lateral restraint are the same for all possible planes of buckling, the true critical load of the element is the value which is calculated from the smallest value of I of the cross-section. If the conditions of lateral restraint are not the same for all buckling planes the plane for which the ratio of L/I is greatest determines the critical load.

The property of the cross-section of a compressive element which is normally used to gauge the critical load is not in fact the second moment of area but a related quantity called the radius of gyration. This is defined by the equation,

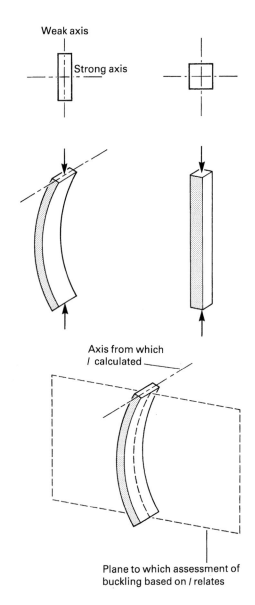

Fig. A2.4 If the conditions of restraint are the same in all planes a strut will buckle about its weakest axis. Thus, the element with the rectangular cross-section is less stable in compression than that with the square cross-section of the same total area.

$$I = r^2 A \qquad (A2.8)$$

where: r = radius of gyration
$ A$ = area of cross-section

The radius of gyration is therefore given by,

$$r = \sqrt{I/A} \qquad (A2.9)$$

The substitution of $I = r^2 A$ is normally made in the Euler buckling formula which then becomes,

$$P_{cr} = \pi^2 E[r/L]^2$$

This can be rearranged into the form,

$$P_{cr}/A = \pi^2 E[r/L]^2$$

$$f_{cr} = \pi^2 E[r/L]^2 \qquad (A2.10)$$

The introduction of $^2 A$ into the formula instead of I therefore allows the critical load to be expressed in terms of a critical average stress. This is a more convenient form for use in design.

Design of compression elements to resist buckling
The relationship between critical average stress and slenderness ratio, which is expressed in equation (A2.10), is shown in the form of a sketch graph in Fig. A2.5. This can be used as the basis of a design method for compression elements.

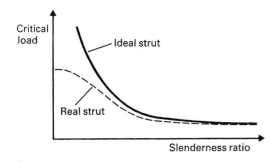

Fig. A2.5 The relationship between critical load and slenderness ratio. The more slender the element the lower is the critical load. The graph shows the relationship as predicted by the Euler formula (ideal strut) and gives an indication of the type of relationship which occurs in practice and which can be derived empirically. The discrepancy is due to the invalidity in practice of one of the assumptions on which the Euler formula is based, namely that all of the material is stressed within the elastic range.

The design of a compression element is a matter of determining a cross-section whose size and shape are such that its buckling strength is greater than the compressive load which will be applied to it. The following procedure can be used in conjunction with versions of Fig. A2.5 for particular materials to achieve this.

1 The compressive force to be carried is determined from the analysis of the structure and the effective length of the strut judged from its actual length and proposed end conditions (see below for an explanation of effective length and the importance of considering end conditions).
2 A trial size and shape of cross-section are selected.
3 From the properties of the trial cross-section and the effective length of the strut the slenderness ratio is calculated and the Euler formula (or graph) is used to calculate the magnitude of the average compressive stress at the critical load value. This is the value of the compressive stress which must not be exceeded if buckling failure is to be avoided.
4 The magnitude of the average compressive stress which will actually occur in the strut is calculated from the applied compressive force and the cross-sectional area of the trial section.
5 If the relationship between the actual stress and the critical stress is not satisfactory the properties of the cross-section are amended and the above sequence repeated. In the interests of safety and economy it is desirable to achieve a cross-section which results in the actual stress being slightly smaller than the critical stress but not excessively so.

The cyclic process is continued until a satisfactory cross-section is achieved.

Slenderness ratio
The critical load of a compressive element is affected by its length and the fact that the length term L appears in the lower part of the

255

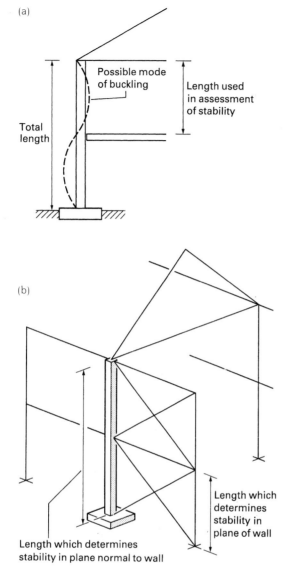

(a)

Possible mode
of buckling

Length used
in assessment
of stability

Total
length

(b)

Length which
determines
stability in
plane of wall

Length which determines
stability in plane normal to wall

Fig. A2.6 The slenderness ratio of an element depends
on the conditions of lateral restraint. Two examples are
shown here.
(a) Assuming that adequate restraint is provided at roof
level and by the intermediate floor, the slenderness ratio
of the column is based on the storey height for the assessment of stability in the plane of the cross-section of the
building.
(b) The lengths on which the slenderness ratios of the
columns in this skeleton frame are based depend on the
plane of buckling under consideration. In the plane of the
wall it is the distance between the points at which lateral
movement is restrained by the cross-bracing. In the cross-
sectional plane of the building it is the height of the
column.

Euler formula indicates that, as would be
expected, the longer an element is the smaller
is the load which it can safely carry in
compression. For a particular element the
length which must be used in the formula is
the distance between the points at which it is
restrained against lateral movement, which is
frequently different from the total length of the
element. Another factor which is significant is
that, in real structure, the conditions of lateral
restraint are frequently different for different
planes of buckling (Fig. A2.6). In such cases the
value which is used for length in the buckling
formula must be compatible with the radius of
gyration which is used as both values must
apply to the same plane of buckling.

A number of different slenderness ratios can
normally be calculated for a particular element
depending on the conditions of lateral
restraint which are provided and on the shape
of the cross-section. Each refers to a particular
plane of buckling – the plane which is normal
to the axis from which the radius of gyration
was calculated. The highest slenderness ratio
is the one which determines the buckling
strength.

The concept of effective length
The characteristics of its end conditions affect
the critical load of a structural element. An
element which has its ends fully restrained
against rotation will carry a higher compressive
load before buckling than one whose ends are
hinged, because the buckled shape of the
fixed-ended element is more complex than
that of one with hinged ends and a greater
load is required to force the element into this
shape. An element with one end fixed and the
other end hinged has a simpler buckled shape
than the element which is fixed at both ends
and its buckling strength is therefore inter-
mediated between the other two cases.

The Euler buckling formula applies only to
compressive elements which are hinged at
both ends but it is possible to use it for
elements with different end conditions by
employing the concept of *effective length*. The
effective length of an element which does not
have hinged end conditions at both ends is the

same as the actual length of the hinge-ended element which has the same buckling strength and which is identical to it in every other respect. Thus the effective length of an element with hinged ends is the same as its actual length. That of an element with fixed ends is approximately 0.5 times its actual length and that of an element with one end fixed and the other completely free is twice its actual length (Fig. A2.7).

Limitations of the Euler formula
Because the assumptions on which it is based are not strictly valid, the Euler formula does not predict accurately the critical load of real elements. In the Euler analysis it is assumed that the critical load value is achieved while the stress in the element is within the elastic range of the material and that instability is due to the inability of the bending stress to resist the bending moment which is caused by the eccentricity which is present. It is not necessary for the stress in the material to pass the elastic limit for this to be possible. The elastic limit of the material would of course eventually be exceeded when the element failed but is not exceeded in the initial stages of the failure. The element does not fail therefore due to yielding of the most heavily stressed part, but because the disturbing couple is greater than the restoring couple while the stress is within the elastic range. The system is unstable, in other words. The phenomenon is known as elastic buckling and it occurs in real structures only if they are very slender.

In most real structures the failure mechanism is slightly different because when a real element is subjected to an increasing amount of compressive load the maximum stress in the cross-section, which of course is the sum of the axial stress and the compressive bending stress, usually becomes greater than the yield stress of the material before the Euler critical load is reached (an indication of the relationship between 'ideal' and 'real' behaviour is given in Fig. A2.5). This causes a sudden increase in the lateral deflection which initiates buckling when the load is less than the Euler critical value. Most real elements there-

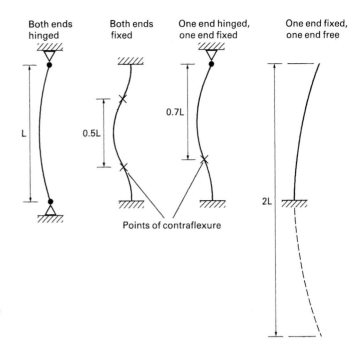

Fig. A2.7 The concept of *effective length*. The effective length of a compression element depends on the end conditions. Effective lengths for different end conditions are illustrated here. (Note that these are theoretical values. The values which are used in practice are different for different materials and will be found in design codes.)

fore fail at load levels which are smaller than the Euler critical load and the less slender the element the greater is the discrepancy between its true failure load and the failure load which is predicted by the Euler formula. The extent of the discrepancy also depends on the type of structure of which the element forms part. The type of material which is used is a particularly influential factor – the behaviour of a masonry pier, for example, is significantly different from that of a steel column.

In practice, therefore, while the design of compressive elements is based on procedures which are similar to the one which has been outlined in connection with the Euler formula, the precise details of the procedures are different for different structural materials. The sequence of operations is broadly the same as the one given above; a cyclic process is used to arrive at a suitable size and shape for the element's cross-section. The permissible stress values which are used in practice are not based on the Euler formula, however. For some materials they are derived from equations which are similar to, but more sophisticated than, the Euler formula while in others they are based almost entirely on experimental data. The practising designer is not normally concerned with the derivation of permissible levels of stress, however, as recommended values can normally be found in codes of practice. They are usually presented in the form of tables or graphs giving permissible values of average compressive stress for different values of slenderness ratio.

The procedures which are used to assess slenderness ratio and effective length tend also to be different for different materials.

A2.4 Steel structures

A2.4.1 Introduction

Steel is used principally in skeleton-frame-type structures so the principal types of element are columns, beams and triangulated girders. The structures are normally hinge-jointed which makes approximate analysis straightforward. Permissible stresses for steel are given in Table A2.4.[2]

A2.4.2 Beams[3]

The most common types of these are floor beams in multi-storey frames, for which the I-section is suitable, or secondary elements, such as purlins or cladding rails, for which smaller sections such as the channel are normally used.

A good approximation to the size required for a beam is given by the equation:

$$Z_{req} = M/f_{pb} \qquad (A2.11)$$

Z_{req} is the required modulus of section and provides the basis for selecting a suitable size of cross-section from the available ranges specified in Table 3.2. If a suitable section size cannot be found from Table 3.2 another type of section will be required (refer to manufacturers' tables). It is normal to select the lightest cross-section which will provide the required section modulus.

M is the maximum applied bending moment determined from the structural analysis. This is determined either from first principles, using the 'imaginary cut' technique, or from the relevant formula in Table A2.3. The load pattern is determined from the area of floor or roof which the beam supports or by estimating the point loads which will be applied to it from other skeleton elements which it supports.

f_{pb} is the appropriate value of permissible bending stress selected from Table A2.4.[3]

The above procedure will give a reasonably accurate prediction of the size required for an element which is subjected to bending moment. A more accurate prediction is obtained if the shear stress and deflection are also checked and the section size adjusted if necessary.

The average shear stress is given by:

$$v_{av} = V/A_{av}$$

v_{av} should not be greater than the relevant value given in Table A2.3. V is the maximum value of the shear force in the beam. This will normally occur close to the supports. A_{av} is the area of the cross-section which resists shear. In the case of steel sections this is the area of the web (the total depth of the section multiplied by the web thickness).

---

2 These should only be used for preliminary sizing of elements to test the feasibility of a proposed structure. They should not be used for final element-sizing calculations.

3 No allowance is made in this procedure for secondary effects such as the compression instability of thin flanges or webs – see Draycott, *Structural Elements Design Manual* (Butterworth-Heinemann, Oxford, 1990). It must therefore be used only for preliminary element sizing. In most architectural structures the parts of elements which are likely to buckle due to local compression are adequately restrained laterally. The estimates obtained from the procedure are therefore likely to be reasonably accurate in most cases.

Table A2.4   Basic permissible stresses for steel [suitable for preliminary sizing of elements only. Not to be used for final design calculations] (BS 449)

Stress type	Basic permissible stress (Grade 43 Steel)
Tension	155 N/mm
Compression	155 N/mm
Bending	165 N/mm
Shear (average)	100 N/mm
Bearing	190 N/mm

Shear stress is only likely to be critical in cases of very heavy loading on relatively short spans.

The deflection of the beam can be checked using the relevant formula from Table A2.3. The critical requirement is that the maximum deflection under the action of imposed load only should not exceed span/360. It is possible for this limit to be exceeded even if the section selected is adequately strong, especially if the load is relatively light and the span long. Where this occurs a larger size of section than that required to satisfy the strength criterion must be selected.

A2.4.3 Columns

Unless there is reason to believe that they will carry a significant amount of bending moment columns should be regarded as axially loaded for the purpose of approximate sizing. For the approximate sizing of steel structures a trial-and-error procedure similar to that outlined in Section A2.3.4 is used and consists of:

1   The selection of a trial cross-section. If the column is axially loaded and the conditions of lateral restraint are the same for all potential buckling planes, the best section shapes are those which have similar bending strength about all their principal axes. The H-shaped universal column (Table 3.3) or square or circular hollow sections therefore perform best in this situation. Section shapes in which the bending strength about one principal axis is significantly greater than about the other, such as the I-shaped universal beam, should be used if the column is subjected to a combination of bending and axial load or if the conditions of lateral restraint are significantly different for different planes.

A preliminary estimate of the size of section which will be required can be obtained by dividing the axial load by an estimate of the final value of the permissible stress. Unless the column is very lightly loaded this will normally be in the upper third of Table A2.5.

2   The calculation of the slenderness ratio of the trial column. As was discussed in Section A2.3.4, a slenderness ratio applies to a particular plane of potential buckling. If the trial cross-section is not symmetrical and/or the conditions of lateral restraint of the column are different in different potential buckling planes the slenderness ratios for these planes, will also be different. The permissible compressive stress is determined by the highest value of slenderness ratio which is calculated.

For a given potential plane of buckling the slenderness ratio is calculated from:

slenderness ratio $= L_e/r$

L_e is the effective length of the column for the plane concerned. This is determined principally by the distance between points at which the column is restrained against lateral movement (normally the storey height of the building). It is affected by the conditions of restraint at these locations, however, and in particular with whether or not any restraint is provided against rotation. This must be assessed and the effective length adjusted in accordance with Table A2.6. In most steel frameworks the effective length is between 0.7 and 1.0 of the distance between lateral restraints.

r is the radius of gyration of the cross-section about the axis which is at right angles to the potential plane of buckling. Radii of gyration of cross-sections are given in Steel Section Tables (see Tables 3.2 and 3.3).

259

Table A2.5a Permissible stresses for steel compressive elements (after BS 449)

L/r	p_c (N/mm²) for grade 43 steel									
	0	1	2	3	4	5	6	7	8	9
0	155	155	154	154	153	153	153	152	152	151
10	151	151	150	150	149	149	148	148	148	147
20	147	146	146	146	145	145	144	144	144	143
30	143	142	142	142	141	141	141	140	140	139
40	139	138	138	137	137	136	136	136	135	134
50	133	133	132	131	130	130	129	128	127	126
60	126	125	124	123	122	121	120	119	118	117
70	115	114	113	112	111	110	108	107	106	105
80	104	102	101	100	99	97	96	95	94	92
90	91	90	89	87	86	85	84	83	81	80
100	79	78	77	76	75	74	73	72	71	70
110	69	68	67	66	65	64	63	62	61	61
120	60	59	58	57	56	56	55	54	53	53
130	52	51	51	50	49	49	48	48	47	46
140	46	45	45	44	43	43	42	42	41	41
150	40	40	39	39	38	38	38	37	37	36
160	36	35	35	35	34	34	33	33	33	32
170	32	32	31	31	31	30	30	30	29	29
180	29	28	28	28	28	27	27	27	26	26
190	26	26	25	25	25	25	24	24	24	24
200	24	23	23	23	23	22	22	22	22	22
210	21	21	21	21	21	20	20	20	20	20
220	20	19	19	19	19	19	19	18	18	18
230	18	18	18	18	17	17	17	17	17	17
240	17	16	16	16	16	16	16	16	16	15
250	15									
300	11									
350	8									

Intermediate values may be obtained by linear interpolation.

NOTE. For material over 40 mm thick, other than rolled I-beams or channels, and for universal columns of thicknesses exceeding 40 mm, the limiting stress is 140 N/mm².

3 The permissible compressive stress is obtained from Table A2.5, based on the highest of the slenderness ratio value calculated in 2.

4 The actual compressive stress which will occur is calculated from the applied load and the area of the trial cross-section (given in the Steel Section Tables) and compared with the permissible value. The actual value should be equal to or slightly less than the permissible stress. If it is not, a new trial section is selected and the above sequence repeated.

Table A2.5b Permissible stresses for steel compressive elements (*continued*)

L/r	p_c (N/mm²) for grade 50 steel									
	0	1	2	3	4	5	6	7	8	9
0	215	214	214	213	213	212	212	211	211	210
10	210	209	209	208	208	207	207	206	206	205
20	205	204	204	203	203	202	202	201	201	200
30	200	199	199	198	197	197	196	196	195	194
40	193	193	192	191	190	189	188	187	186	185
50	184	183	181	180	179	177	176	174	173	171
60	169	168	166	164	162	160	158	156	154	152
70	150	148	146	144	142	140	138	135	133	131
80	129	127	125	123	121	119	117	115	113	111
90	109	107	106	104	102	100	99	97	95	94
100	92	91	89	88	86	85	84	82	81	80
110	78	77	76	75	74	72	71	70	69	68
120	67	66	65	64	63	62	61	60	60	59
130	58	57	56	55	55	54	53	52	52	51
140	50	50	49	48	48	47	47	46	45	45
150	44	44	43	43	42	42	41	41	40	40
160	39	39	38	38	37	37	36	36	36	35
170	35	34	34	34	33	33	33	32	32	31
180	31	31	30	30	30	30	29	29	29	28
190	28	28	27	27	27	27	26	26	26	26
200	25	25	25	25	24	24	24	24	23	23
210	23	23	23	22	22	22	22	22	21	21
220	21	21	21	20	20	20	20	20	20	19
230	19	19	19	19	19	18	18	18	18	18
240	18	18	17	17	17	17	17	17	17	16
250	16									
300	11									
350	8									

Intermediate values may be obtained by linear interpolation.

A2.4.4 Triangulated elements

Triangulated girders and trusses can have a variety of overall forms. The most commonly used are the pitched truss and the parallel chord truss. Pitched trusses normally have a pitch angle of between 30° and 45°. This determines their overall depth. The depth of parallel chord trusses in relation to their span can vary widely. Principal structural elements which carry substantial areas of floor or roof normally have span/depth ratios in the range 12 to 14. Lightweight triangulated steel joists are much less deep with span/depth ratios in the range 20 to 30.

Table A2.6 Effective lengths of steel compressive elements for different conditions of end restraint (after BS 449)

Conditions of end restraint	Effective length of element (L = distance between restraints)
Effectively held in position and restrained in direction at both ends	0.7L
Effectively held in position at both ends and restrained in direction at one end	0.85L
Effectively held in position at both ends but not restrained in direction	L
Effectively held in position and restrained in direction at one end and at the other partially restrained in direction but not held in position	1.5L
Effectively held in position and restrained in direction at one end but not held in position or restrained in direction at the other end	2.0L

The internal geometry of triangulated girders is arranged so that very small internal angles (less then 30°) are avoided. The ideal arrangement is one of equilateral triangles.

The individual sub-elements of triangulated structures can be regarded as carrying either pure axial tension or pure axial compression. The magnitudes of the internal forces on any particular sub-element can be determined by using the *method of sections*. This is a version of the 'imaginary cut' technique. Once the magnitudes of the axial forces in the sub-elements are known the size is determined by using the techniques outlined in Sections A2.4.2 or A2.4.3.

Axial internal forces are best resisted by cross-section shapes which are symmetrical, such as circular or square hollow sections. Angle and channel sections are also commonly used.

The feasibility of a particular structural proposal can be checked by sizing only the most heavily loaded of the sub-elements. In parallel-chord arrangements these are the horizontal sub-elements at mid-span and the inclined or vertical sub-elements close to the supports.

A2.4.5 Elements subjected to combined stress

The basic rule for elements which are subjected simultaneously to more than one type of internal force is that the following equation must be satisfied:

$$f_a/f_{pa} + f_b/f_{pb} < 1 \qquad (A2.12)$$

where: f_a = actual axial stress
  f_{pa} = permissible axial stress
  f_b = actual bending stress
  f_{pb} = permissible bending stress

The equation must be satisfied at all locations in the element. If the internal forces vary along the length of the element it may be necessary to check that the equation is satisfied at more than one location.

A2.5 Reinforced concrete structures

In reinforced concrete structures the elements are subjected to primary load actions which are either of the bending type (beams and slabs) or of the axial compression type (columns and walls). In the case of bending-type elements there are two considerations which affect the overall sizes of the cross-sections which must be adopted – the provision of adequate bending strength and the prevention of excessive deflection. In the case of compressive elements, the prevention of buckling-type failure is an important consideration. This is dependent on slenderness. Often reinforced concrete walls or columns are relatively short, however, and the overall dimensions which are adopted can be determined by the need to prevent the level of compressive stress from becoming excessive rather than the need to maintain the slenderness ratio low enough to maintain adequate resistance to buckling.

The strength of a reinforced concrete element is determined principally by the overall size of its cross-section, by the strength of the constituent concrete and by the quantity and location within the cross-section of the reinforcement which is positioned within it. The strength of concrete can vary widely (typically between 25 N/mm^2 and 70 N/mm^2) depending on the mix proportions and water/cement ratio which are specified. It is therefore possible, by manipulating the strength of the concrete and the amount of reinforcement which is present, to produce a range of strengths within a given overall size of cross-section. For this reason the overall dimensions of elements tend to be determined by other criteria than the need to provide adequate strength. These are the need to limit deflection in the case of bending-type elements and the need to avoid excessive slenderness in the case of axially loaded elements. The 'rules of thumb' given in Table 4.1 provide reasonably realistic approximations to the sizes which will be required for reinforced concrete elements.

A2.6 Masonry structures

As with reinforced concrete, the basic strength of masonry can vary over a fairly wide range depending on the strengths of the constituent bricks or blocks and the mix proportions of the mortar which is used. For this reason the overall thickness of walls and columns tends to be determined by considerations of limiting the slenderness ratio or constructional convenience rather than by the maximum load which is carried. Basic masonry elements may therefore be sized approximately from the data given in Tables 5.1 and 5.3.

A2.7 Timber structures

A2.7.1 Introduction

Rigorous structural design calculations for timber are complex because the carrying capacity of timber elements can be affected by a large number of factors. Included in these are the duration of the load, the moisture content of the timber, the extent to which load sharing between elements is possible, the overall dimensions of the element, the direction of the load in relation to that of the grain, and a number of other effects. These are taken into account in rigorous element-sizing calculations by adjusting the value which is used for the permissible stress to suit the individual circumstances of the structure. The permissible stress which is used in a particular case is the basic allowable stress for the species involved multiplied by one or more stress-modification factors. The exact procedure which is used to evaluate the permissible stress in a particular case depends on the type of element under consideration (i.e. on whether it is solid or laminated, etc.) and on the nature of the internal force which it will carry. These procedures will not be explained in detail here.

The fact that so many factors can affect the strength of a timber structure means that simplified approximate sizing calculations give a slightly less reliable indication of the final sizes which will be required than is the case with equivalent methods for other materials. Particular care is required when the limits of the spans in which a particular component is normally used are approached. This is especially the case with bolted trusses in which the viability of a proposed arrangement is likely to be determined by the feasibility of the joints. Where the limits of normal practice are approached, the feasibility of a proposal can only be tested by carrying out rigorous sizing calculations.

A2.7.2 Tension elements

The area of cross-section required can be determined from:

$$A_{req} = P/f_{ap} \qquad \text{(A2.13)}$$

where:

A_{req} = area of cross-section required. If the end connection is of the bolted type the cross-section selected should be increased by 20%

P = applied axial load

f_{ap} = permissible axial stress. This will depend on the species of timber. Typical values are:

softwood	4 N/mm^2
laminated timber	6 N/mm^2
hardwood	9 N/mm^2

A2.7.3 Bending elements

Solid rectangular sections

The section modulus required is found from:

$$Z_{req} = M/f_{pb} \qquad \text{(A2.14)}$$

where:

Z_{req} = section modulus required ($bd^2/6$ for a rectangular cross-section)

M = maximum applied bending moment

f_{pb} = permissible bending stress. Typical values for this are,

softwood	6 N/mm^2
laminated timber	12 N/mm^2
hardwood	15 N/mm^2

The deflection is more likely to be critical than with other materials and should be checked. Because most timber beams carry distributed loads, the requirement for adequate stiffness is satisfied if the following relationship is complied with:

$$I_{req} = 4.34WL^2/E \qquad \text{(A2.15)}$$

where:

I_{req} = second moment of area of cross-section ($bd^3/12$ for a rectangular cross-section)

W = total applied load

L = span

E = modulus of elasticity of timber. Typical values for this are:

softwood	8000 N/mm^2
laminated timber	10000 N/mm^2
hardwood	15000 N/mm^2

A2.7.4 Compressive elements

Timber columns and other compressive elements must be sized from the trial-and-error procedure outlined in Section A2.4.3. The permissible stress is determined by the slenderness ratio. Typical values of permissible stress are given in Table A2.7.

Table A2.7 Typical permissible compressive stresses in timber

Slenderness ratio L_e/r	f_{pc} (N/mm^2) Softwood	Laminated timber	Hardwood
0	7	8	16
5	6.83	7.80	15.60
10	6.66	7.61	15.22
20	6.27	7.17	14.37
30	5.79	6.62	13.23
40	5.15	5.88	11.76
50	4.35	4.97	9.94
60	3.54	4.05	8.10
70	2.86	3.26	6.53
80	2.31	2.64	5.28
90	1.90	2.17	4.34
100	1.58	1.80	3.60
140	0.85	0.97	1.94
160	0.66	0.75	1.50
200	0.43	0.49	0.97

For rectangular cross-sections, $r = 0.288t$

t = the smaller of the cross-section dimensions.

Index